Praise for *SICK! Curious Tal*
Share with

Understanding that most of the infectious diseases and pests that affect animals also affect people is something that veterinary students learn early in their academic careers. *Sick!* brilliantly illustrates the inextricable link between the health of animals and people through captivating, real life stories. This link is aptly labeled "One Health" and underlines the need for closer collaboration between all of the health care professions.

Drs. Stone and Dewey have my admiration and gratitude for bringing "One Health" to life for all of us.

Dr. W. Ron DeHaven, Chief Executive Officer
American Veterinary Medical Association

Perfectly demonstrates the role of veterinarians in protecting human health! If you want to know about the intimate link between the health of humans, animals, and the environment - and how our futures are irrevocably interdependent - this book is for you. From cats to rats, birds to beasts, and mice to men, these stories cover the breadth and depth of the complex world of infectious disease. Read it to prevent getting Sick!

Dr. Andrew T. Maccabe, Executive Director
Association of American Veterinary Medical Colleges

Being an expert on the human-animal health connection, I'm painfully aware of the two edges of the sword ... animals that can make us happy and healthy can also make us very sick. From mundane problems like ringworm ... to deadly outbreaks of zoonotic diseases such as Ebola virus ..., "Sick!" teaches valuable lessons in what these diseases are and how to minimize the risk and maximize the benefits we receive from sharing our lives with animals. I highly recommend this book.

Dr. Marty Becker "America's Veterinarian"
Veterinarian for Good Morning America and The Dr. Oz Show

"Sick! is a captivating, engaging, and stimulating exploration of the many connections between human and animal health and disease. The storytelling is irresistible and the illustrations help to truly clarify the sometimes very complex relationships between hosts, pathogens and ecosystems. Sick! should be read by everyone who enjoys engaging storytelling and is interested in the vital linkages between humans, animals and the ecosystem we all inhabit."

Barbara Natterson-Horowitz, M.D.
Professor of Medicine and author of New York Times Bestseller, *Zoobiquity*

"I have thoroughly enjoyed reading this book. It has reminded me not only of all the amazing aspects of veterinary medicine, but also the role that veterinarians play in human health, community outreach, and zoonoses.

Congratulations on compiling such a diverse and interesting assortment of veterinary tales."

Jim Berry
Past President, Canadian Veterinary Medical Association

"I found this book to be a thoroughly engaging compilation of stories. Some anecdotal, some works of fiction and some reflections of personal experiences, all worth reading.

The real value of the book of course, beyond the warmth and unique nature of the individual stories, is the common bond that runs through them all which is how the health of humans and animals is so intimately connected in the bio-sphere that we share which, in itself, is an important story to be told.

Who knew that so many accomplish veterinarians and eminent scientists could write such poignant literature with humour, sincerity, and empathy? It places the rest to myth that scientists are not good communicators.

This is a book that belongs in the libraries of every school and every health professional, and also in the waiting rooms of every doctor's office, whether human or veterinarian"

Dr. Bernard Vallat
Director General, World Organization for Animal Health (OIE)

Sick! Curious Tales of Pests & Parasites We Share with Animals

Elizabeth Arnold Stone and Cate Dewey, Editors

Ontario Veterinary College
University of Guelph
Guelph, Ontario

For information about permission to reproduce selections from this book, write to Office of the Dean, Ontario Veterinary College, University of Guelph, 50 Stone Road East, Guelph, ON, Canada, N1G 2W1.

ovcdean@uoguelph.ca www.ovc.uoguelph.ca

"Dog Parasites in the Land of the Gods", a selection from *The Chickens Fight Back: Pandemic Panics and Deadly Diseases That Jump from Animals to Humans* by David Waltner-Toews, is adapted courtesy of Greystone Books.

Printed in the United States of America. Published simultaneously in Canada.

The text of this book is set in Garamond.

Library and Archives Canada Cataloguing in Publication

Sick! : curious tales of pests & parasites we share with animals / Elizabeth Arnold Stone and Cate Dewey, editors

Includes bibliographical references and index.
ISBN 978-0-88955-609-6 (pbk.)

1. Zoonoses. 2. Parasites. I. Dewey, Catherine E.
II. Stone, Elizabeth Arnold
III. Ontario Veterinary College. issuing body

SF740.S52 2014   636.089'6959   C2014-901315-9

Ontario Veterinary College, University of Guelph, 50 Stone Road East, Guelph, ON, Canada N1G 2W1

www.ovc.uoguelph.ca

Illustrations: Tony Linka

Cover design: Jane Dawkins. Cover art: Char Hoyt

3 4 5 6 7 8 9 0

CONTENTS

LIST OF FIGURES

Introduction

People and animals have co-existed throughout history. Pets sleep in the bedroom or even on the bed. Farmers work every day with their livestock. People venture into the wilds, exploring fields and forest; they hike and canoe, stopping for a drink from the river or lake. People even keep wildlife as pets – raccoons, monkeys, reptiles – and the list goes on. Animals are so important to all of us. For some people, animals provide food and transportation; for others, animals may improve family nutrition and health; and for others, animals may bring joy and companionship.

At the same time, we are united, humans and animals, by the company we keep – sharing the parasites and microbes that find their homes within and upon us. Many of these creatures cause us no harm; in fact, some are necessary for us to live. We couldn't digest our food if it weren't for the "good bacteria" living in our gut. Nevertheless, even these bacteria can cause disease when they are in the wrong place at the wrong time.

For this collection of "curious tales of pests and parasites", we asked veterinarians to share their stories, their experiences working with "zoonoses". These zoonotic diseases are caused by the infectious organisms we share with animals – and over 60% of all infectious diseases fall into this category. Our authors write about parasites that may not bother the animal but cause serious diseases in people. Other organisms cause diseases in both humans and animals. And many parasites, such as the pig tapeworm, cleverly use both human hosts and animal hosts to complete their life cycle. Usually we hear about animals transmitting disease to people as if it is a one-way street. Veterinarians are acutely aware that this is only half the story. People can also make animals sick - pigs on a farm infected with influenza from a worker who had traveled abroad; captive dolphins with antibiotic resistant infections from their handlers.

Our student veterinarians learn about the health of populations, preventive medicine, parasitology and epidemiology – and they study diseases in many different species. Thus, all veterinarians are grounded in a firm understanding of the complex ecosystem of people and animals and the environments in which they live: the monkey spreading Ebola virus to humans; the caged hamster with mites near the boy's bed; the fish eaten by a seal that is killed by an Inuit hunter; the Kenyan woman dependent on her pig for financing her children's education.

As people encroach on wildlife habitats, the parasites, bacteria and viruses can more easily jump from animal host to human host. And global travel means that zoonoses occurring "far away" could soon be at our doorstep or in our living room. These organisms don't recognize national borders and veterinarians are often on the front line to identify and control these trans-border diseases, such as West Nile virus, Mad Cow Disease and avian flu.

Some of the stories in this collection are humorous; others are gripping; some are sad. Tales of brucellosis and tuberculosis give a historical perspective from the days when veterinarians cared for all creatures great and small. An account of laboratory work to protect Second World War soldiers from botulism foreshadows modern day fears of bioterrorism with anthrax. Tales from around the world remind us of our responsibilities for global health. For example, even though bovine tuberculosis was eradicated in Canada, badgers keep it going in Ireland. Stories about Uganda, Kenya, and Iceland interweave zoonotic diseases with the daily lives of people in these countries. Some diseases described in this book, such as Ebola virus, rabies and sleeping sickness, are terrifying. Others, like flea infestations are more benign – or are they?

Original line drawings illustrate many of the stories to help readers easily grasp the sometimes complicated ways in which people may contract these diseases.

We hope you are as fascinated by these stories as we are. Likely, you will see your environment and the animals around you in a new light. People and animals co-existing: there's a world of mystery and intrigue that lies therein.

– EAS and CD

Too Close for Comfort: Encounters with Giardia in Urban and Rural Communities

Heather Bryan, Maëlle Gouix and Judit Smits

Editor's note: Giardia is a unicellular protozoan parasite that infects people and animals around the globe. While conducting research on wolf-parasite interactions, including giardia, Heather Bryan and Maëlle Gouix ironically encountered a bout of giardia themselves and shared their experiences with their supervisor, Dr. Judit Smits. It turned out Dr. Smits had had her own experience with giardia. Their two stories highlight the challenges associated with detecting, treating, and determining the sources of giardia infections.

Mist swirled around us and muffled the purring engine of the zodiac taking us to shore. As we approached the estuary, the shapes of tall cedars emerged from the shroud. Peering intently through binoculars, it was Maëlle who spotted several pairs of pointy ears sticking up above yellow-tinged sedges. Awestruck, we watched as three wolves briefly looked us in the eyes before disappearing into the surrounding rainforest.

Thrilled by our glimpse of these elusive carnivores, we debarked from the zodiac in hopes of discovering what the wolves had left behind. We were not disappointed; the wolves had made distinct trails through the vegetation that led us to several fresh wolf scats (feces). Though considered waste by some, these scats contained evidence of parasites that would provide insight into the natural relationship between wolves and their parasites.

Over the following three weeks, we continued our search for wolves and their scats on island and mainland regions of the stunning Great Bear Rainforest in coastal British Columbia. The sample collections were part of my graduate program at the University of Saskatchewan and contributed to a long-term wolf monitoring program run by the Raincoast Conservation Foundation.

One of our team members was Dr. Maëlle Gouix, a visiting veterinarian from France working on our project under the supervision of my graduate PhD advisor, Dr. Judit Smits. Among her many talents,

Maëlle has an incredible sense of humour that she used to entertain us with tales of her world travels and veterinary work.

A second goal of our project was to generate discussion about the links among wildlife, dog, and human health. To this end, we offered dog health clinics in several communities where we examined, vaccinated, and de-wormed dogs. We also collected blood and stool samples from dogs to look for evidence of infectious diseases and parasites that might affect the health of dogs and their wild relatives, wolves.

One of my most vivid memories from the clinics occurred in the final community we visited, where we were greeted by four happy, friendly puppies and their mother. In addition to adult people who brought their pets to the clinic, several children came for a visit. Some of the children brought their own pets, while those without pets brought stuffed animals for veterinary attention.

The children followed the examinations closely on their stuffed or real animals. Their favourite part was watching Maëlle take a stool sample. They covered their eyes as she used her gloved hand for the collection, then open it dramatically to reveal . . . oooh, poo! Every time, this would result in a flurry of giggles and shrieks.

We were grateful for the people and dogs we had met and for the generosity of community members who had helped with the clinics. Subsequently, Maëlle and I, along with the rest of our research team, continued collecting wolf scats for several weeks. When we returned to Saskatoon a month later, we forged ahead with analyzing the fecal samples we had collected for the eggs and larvae of worm-type parasites.

It's smelly and tedious work. We extracted eggs and larvae from the feces, transferred them to rectangular glass slides, and examined them through a microscope. For giardia and Cryptosporidium, which are protozoan parasites, a special assay was needed because they are too small to identify with regular light microscopy.

For the protozoal assay, we smeared a small amount of feces on a microscope slide and incubated the smear with a fluorescent antibody designed to target the cysts of giardia and the oocysts of Cryptosporidium. The antibody makes the cysts and oocysts show up as bright green under a special fluorescent microscope.

Examining both sets of slides from the day's fecal samples took hours. By the end of each day, Maëlle and I were exhausted, our eyes tattooed with black rings from pressing our faces against the microscopes. After one of these taxing days, Maëlle mentioned that she had been feeling unwell lately. Her stomach had been bloated and she'd been having diarrhea.

Based on the symptoms, Maëlle already had a strong intuition about what illness she might have. The next day she felt so sick she couldn't come to work. Her doctor prescribed medication to kill protozoa (metronidazole) and she soon felt better.

While Maëlle was recovering, I started to notice similar symptoms. When I walked up the stairs to the lab, I felt completely exhausted. My stomach was bloated with gas and I seemed to use the toilet urgently and more often than normal. I began to wonder if I had the same illness. Or was I was just tired from the fieldwork and long days in the lab?

I decided to find out. From two previous experiences with giardiasis – both caused by my foolish consumption of untreated water abroad and in Canada – I was familiar with the challenges associated with diagnosis. Both times, my stool sample had come back negative and the doctor had been reluctant to treat me. It wasn't until the symptoms persisted for several weeks after my initial visit that I convinced my physician that something was wrong and needed treatment.

Now armed with a better understanding of fecal-based tests for giardia, and knowing that giardia cysts are shed only intermittently in feces, I prepared to convince the doctor that my intestines were indeed inhabited by unwanted visitors.

Accordingly, on my next dash to the toilet, I went armed with a plastic collection container. The next morning, I prepared the day's slides to check for giardia, but this time also added a couple of extra samples from myself. Upon examination of my sample under the microscope, I immediately saw that it contained a sea of small green giardia cysts. In fact, the slide was so bright with green fluorescence that it almost hurt my eyes to look at it. Both horrified and excited by such an obviously positive sample, I called Maëlle to come look at this startling specimen. She had never seen a sample as spectacular as this one.

In contrast to mine, most of the wolf fecal samples contained only a few cysts. In fact, we had to scrutinize each sample carefully so we did not miss the one or two cysts that were present. The low-grade infections in wolves are typical of many wildlife populations, where most individuals have low or no parasitic infections and a few have high levels of infection. And individual wolves infected with intestinal parasites, including giardia, are probably used to living with infections that are maintained at a low level. Indeed, the hosts and parasites likely evolved together over thousands of years and no doubt have struck a balance.

After examining the rest of the day's slides, I headed straight to the doctor. Feeling slightly embarrassed about my self-diagnosis, I explained my symptoms and reasons for suspecting giardia. When I got to the part about examining my own stool, I was surprised that the doctor hardly showed any reaction. Instead of asking for another specimen to confirm the diagnosis, he simply handed me a prescription.

Once we were both feeling better, Maëlle and I began to wonder how we had become infected with giardia. The small, unicellular organism is ubiquitous throughout the world and is the most common parasitic infection in the intestines of humans. When a host ingests a giardia cyst, it travels to the digestive tract and settles in the small intestine. It then transforms into the feeding stage, the trophozoite, attaches to the inner wall of the intestine and divides rapidly, gorging itself on nutrients its host has eaten.

When fully sated, the trophozoite encases itself into a cyst that is temperature resistant and travels to the outside world in its host's feces. The cyst gets into its next host directly from something that has feces on it (e.g., hands) or from contaminated water or food.

Reflecting on the time course of our infections, Maëlle and I ascertained that we likely acquired the giardia during our field research. Indeed, we found evidence of giardia infections in wolf scats, including strains that are infectious to humans. However, only 7% of the 1,558 specimens contained giardia cysts and the number of cysts in most fecal samples was low. Though we noted several dogs' feces with evidence of giardia infections, we took precautionary measures to prevent contact with feces during clinics.

Specifically, in both field and clinic settings, we always wore gloves and frequently washed our hands to protect ourselves. Moreover, the chance that both of us had ingested a piece of infected feces at about the same time seemed unlikely. None of the other eight people helping us with fecal collections reported diarrhea or other symptoms of giardia.

Perhaps Maëlle and I had come into contact with infected water as we splashed through streams during our sample collections. Another possibility is that both Maëlle and I drank tap or well water containing giardia cysts. During our research, Maëlle and I might have missed a boil water advisory in the communities we had visited. In the end, we never could confirm the source of our infections. However, we described the incident to our supervisor, Dr. Smits, and discussed solutions for avoiding future infections. In the process of our consultation, quite coincidentally it turned out, we learned about Dr. Smits' own giardia story.

Here's her account.

Daycare diarrhea, the canoe trip and the pesky parasite

My daughter, little Miss Veronika, loved to go on canoe trips with her Mama and Papa and other big- and medium-sized family friends. Her job, starting as a toddler, was to entertain herself in the middle of the canoe, dragging her arms in the water, using a stick to make splashes and maybe even catch floating plants, or sometimes finding perfect bull rush stems in the marshy landing spots beyond which the tents were being pitched for the nightly camp.

Seven days after one of these wonderful canoe trips, Veronika was at her favourite daycare with her diaper-clad buddies in the toddler group, when she started needing many more diaper changes. She didn't complain much, but would just suddenly have to go. The dream of toilet training was looking less and less likely, but a few days later, after twice daily doses of Kaopectate, everything was alright again . . .

until a week later when the diarrhea started again. As before, she had no fever, no decrease in appetite and no decrease in energy, just mild complaints and increased diaper, clothing and sheet changes.

With my veterinary background, I started thinking about the connection between the off-and-on again diarrhea and the intimate contact with water we had shared with muskrat and beaver. Our canoe trips do, after all, take us through the living spaces of these semi-aquatic mammals. I recalled from lectures during my veterinary studies, that in dogs with giardia, it helped to decrease carbohydrate levels in the diet.

So this child of a veterinarian got the same treatment, but we also took a small sample of her feces to the medical laboratory with a request to look for parasites that commonly cause diarrhea, and in particular, giardia. But the tests came back negative and the laboratory told us confidently that Veronika did not have giardia. I wasn't convinced because I know how difficult giardia is to detect – Veronika's diarrhea had continued. I talked our physician into a prescription of metronidazole (Flagyl), a well-known, effective medication against *Giardia lamblia*.

Several times a day we treated her with the chocolate flavoured but still nasty tasting medicine accompanied by howls of protest, impressive faces, squirming, spitting, cajoling and trickery. But despite all this, the medication was delivered on schedule as required.

Meanwhile a few days after Veronika's treatment had begun, her little friends at the daycare started having diarrhea. We talked to the daycare staff about giardia, the difficulty of diagnosis and how it is spread. The staff tried hard to be extra vigilant with the hand washing and hygiene during diaper changes. But trying to control contact among toddlers proved impossible, and within a few days, several other toddlers had diarrhea. Of course these parents had to visit their family physicians with the children, try to get diagnoses through stool samples, and all the while being reluctant to start their children on medication without a definitive diagnosis.

By the time other families started treatments on their children, our family had struggled through one cycle of medication and Veronika no longer had the parasite or the diarrhea. Two parents we knew well, one

of whom is a physician, ended up with their entire family infected –
which was diagnosed from stool samples - and all had to be treated
with the unpleasant metronidazole.

As luck and bad planning would have it, some other children with
diarrhea had not been positively diagnosed with giardia and were not
treated. They provided the perfect source of re-infection for our
Veronika and other children. At this point, we tried hard to involve
the public health agency to get the Giardiasis under control. The
public health officials told us that we were making much ado about
nothing and were not interested in doing an investigation or talking to
parents about the infection and the best strategies for dealing with it.

Thanks to the fact that we had two parents with medical degrees,
albeit one in veterinary medicine, we managed to convince everyone
that it was critical to get on top of the infection or it could get badly
out of control.

We started again with the demanding treatment regime. This time,
with several parents and the daycare director and staff now fully
convinced of the reality of the risk, we managed to co-ordinate
treatment of affected toddlers, defeat the giardia and regain the sunny
atmosphere of the daycare.

There was no epidemiological investigation to trace the source of
this outbreak, but the temporal sequence of events during the canoeing season, suggests our family, or the other canoeing family could
have had the original exposure. This outbreak occurred in a professional, well-run daycare with only 10 children in the youngest "diaper
group." If the problem had occurred in a larger group, or one without
medical expertise among the parents, , I can only imagine the chaos
and spread of giardia into the families and community that might have
resulted.

Epilogue

As our two experiences with giardia illustrate, giardiasis is prevalent
and borderless, showing up from remote First Nations communities to
urban centres. In addition, while no definitive source was identified in
either of our experiences, animals were likely involved in transmission. Certainly, they were the source of the infections. Wilderness
camping is essentially placing ourselves in wildlife's habitat.

One distinction between the two incidents is the contrast of settings. When the toddlers were infected in the urban setting, resources were readily available for a response, even if some of those involved were slow off the mark. In contrast, even though no definitive link was made to the remote research site, the infection demonstrates the importance of veterinary care for the health of the entire community. As well, it illustrates how crucial it is that residents of Canada's remote communities have access to resources to maintain their health in the face of such threats.

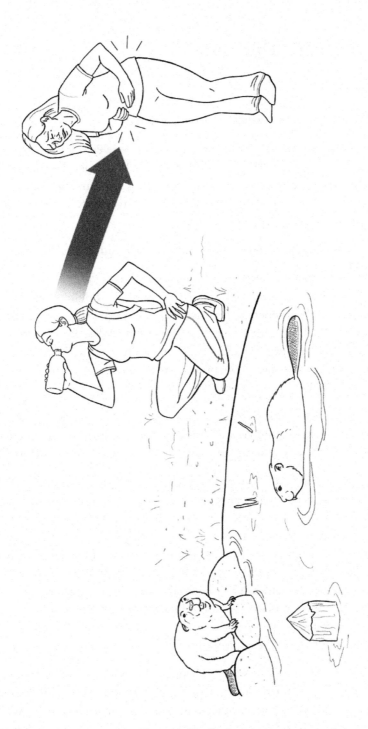

Figure 1 Giardiasis

A Mother's Dilemma

Elizabeth Stone

"Please come immediately and pick up Karen!"

The urgent message from the daycare is one every mother dreads. First, I ask myself, what could be so bad that I must get there immediately? I'm scheduled to do a surgery on a dog in 10 minutes, how can I leave? But of course, right away, I call the daycare to tell them I'm on my way, asking, "What is wrong?"

"Your daughter has impetigo and we don't want the other children to catch it!" is the answer.

When I dropped Karen off a few hours earlier, I didn't notice anything resembling impetigo, but then I'm a veterinary surgeon and don't know much about dermatology for animals, much less for people. Fortunately, my surgery patient isn't yet anesthetized, so I ask the anesthesiologist to cancel the procedure and I rearrange my surgery schedule to do it the next day.

On my way to the daycare, I think, "Why is this happening? We had just returned from vacation the night before, hadn't even had time to pick up the dog and cat from the boarding kennel, and now this!"

Karen, who is three-years-old, is glad to see me in the middle of the day and I'm beginning to look forward to our afternoon together. I ask the teacher to show me the lesions they think are impetigo. On her legs from her ankles to her knees are small, red dots. After taking one look, I cavalierly explain that those spots aren't impetigo, they're just flea bites. My goodness, the look I get – "*Just* flea bites?! Do you mean your house has fleas?" – no doubt implying we must live in a slovenly, unkempt place. No doubt there is an element of truth in what she thinks about our busy household, but basically our house is clean enough not to be a health threat.

Let me explain. This scenario happened in the mid-1980s, before the advent of effective flea control. Thus, any household with dogs in the Southern USA (and in other warm, moist climates) had fleas. We could "bomb" the house with insecticide, spread Sevin™ dust on the carpets and spray the yard, but fleas were still a problem.

Despite its name, the most common household flea in North America is the cat flea (*Ctenocephalides felis*), which prefers the taste of cats and dogs and "jumps" at the chance to burrow into their fur or bedding. If a flea does get on a person, it rarely stays long enough to be carried to another location or another person. As long as a person bathes frequently and washes their hair, there is no chance for a flea to lay eggs and set up housekeeping. And even as a mom working outside the home, I did make sure Karen had a bath every day!

Then why did Karen get so many bites? The answer comes from the flea life cycle. When we went on vacation, the fleas stayed behind, where they busily laid eggs on the pets' beds. Tiny, worm-like larvae hatched from the eggs, fed on the stool of the adult fleas, molted twice and formed a cocoon. Inside the cocoon, a final molt turned the larva into an adult flea. The adult fleas were waiting inside the cocoon for food to arrive (and they can survive for a year like this without food or water).

When our family returned home from vacation, the pets were still at the boarding kennel. Consequently, we were the next best source of food. When the fleas felt the vibrations we made in the floor (or maybe even sensed an increase in carbon dioxide in the air from our breath), they jumped out from their cocoons. That morning as I was getting ready for work, I had noticed a few fleas on my pantyhose (yes, the 1980s was a time when we wore such things), but didn't think much about it. A flea can jump up to eight to 10 inches (150 times its own size), so they often land on a person's ankles or legs.

The body of a flea is tough and you can't kill them just by squeezing them between your fingers. What works best is rolling them between your thumb and finger and then crushing them with your fingernail. It's also said that fleas can be drowned in water, e.g., the toilet, as long as you can get the flea into the water before it escapes! As I write this, I realize how adept we were at dealing with these little pests. I doubt that mothers in this 21st century can so easily recite 'the best way to kill a flea'!

In 1665 Robert Hooke, who was one of the first to use microscopes for scientific study, described the flea:

"It has a small proboscis, or probe, that seems to consist of a tube and a tongue or sucker, which I have perceiv'd him to slip in and out. Besides these, it has also two biters, which are somewhat like those of an Ant..; these were shap'd very like the blades of a pair of round top'd Scizers, and were opened and shut just after the same manner; with these Instruments does this little busie Creature bite and pierce the skin, and suck out the blood of an Animal, leaving the skin inflamed with a small round red spot."

Flea bites may not show up right away (or at least that was my mother-guilt reducing theory), and anyway, in my mind, flea bites would not be a reason for me to keep Karen from daycare. However, for people who don't know about fleas and their bites, they may mean the spread of disease.

Why wasn't I worried about my daughter getting a disease from our household fleas? The cat flea doesn't carry any serious diseases, but it can host the dog tapeworm, *Dipylidium caninum*, which can infect both dogs and cats. I realize this can be confusing – cat fleas on dogs and dog tapeworms in cats! Occasionally, if a child ingests a flea that has a tapeworm, the tapeworm larva will escape from the flea and develops into a tapeworm in the bowel (intestine) of the child. This species of tapeworm doesn't cause obvious symptoms; usually it is noticed when the parent or caregiver sees what looks like a grain of rice on the child's buttocks or in their stool.

But what about other fleas? With more than 2,000 described species of fleas in the world, it's not surprising that some of them can spread disease. Probably the most infamous disease is bubonic plague, the cause of "Black Death" in the 1340s. This infectious disease was spread by rat fleas (*Xenopsylla cheopsis*) to people.

And now on to the second part of this story…

My sister, Talitha, minister of the United Church of Santa Fe in New Mexico, was on the phone and she sounded worried. Two friends, John and Linda, had flown to New York City (NYC) and within two days had become so sick they were admitted to a hospital. Thinking about the kinds of things that make people sick while traveling, I asked: "Was it food poisoning or the flu?"

Soon we learned the answer – from a headline and story in *The New York Times*!

New Mexico Man, 53, Is Seriously Ill With Plague at Beth Israel

A 53-year-old New Mexico man was in critical condition last night at Beth Israel Medical Center with bubonic plague, the rare and deadly disease that once decimated Europe, health officials said.

His wife, a 47-year-old woman, remains under observation at Beth Israel as tests for the disease are conducted.

The story, published on November 7, 2002, continued,

At a news conference last night, Thomas R. Frieden, the city's health commissioner, said that both victims were residents of the Santa Fe area who had arrived in New York City on a vacation on Nov. 1. He said that they became ill on Sunday. The timing provided evidence that the man, and perhaps the woman, became infected before arriving in the city, since the incubation period before symptoms of plague become apparent is two to seven days.

Dr. Frieden said that New York City health officials had been in contact with the medical authorities in New Mexico, and had obtained more evidence that the infection most likely occurred in that state.

The New Mexico officials found evidence in July that a dead wood rat on property owned by the couple had been carrying fleas infected by the bubonic plague, he said.

"We are confident the exposure occurred in New Mexico," Dr. Frieden said. "There is no risk to New Yorkers from the individuals who are being evaluated for plague."

Other news media outlets were not so circumspect. Headlines in the New York Daily News screamed "City Plague Case 1st in 100 Years!" Fox News flashed an onscreen banner that intoned "Black Death" while the newscaster reported that two tourists had caught the bubonic plague.

With the recent memory of the September 11, 2001 attacks on New York City and the subsequent anthrax scare, bioterrorism was an initial concern. Were the husband and wife the first two victims of an attack? Were these two people chosen by bioterrorists to carry plague into New York City? Quickly, public health officials reassured people that this city of eight million was in no danger. They emphasized that the couple became ill within 48 hours of arriving, thus their exposure most likely occurred before they left New Mexico.

Known as Black Death in the Middle Ages, the bubonic plague has killed more than 30 million people throughout history. It is a bacterial disease of rodents transmitted to humans through the bites of infected fleas, especially the rat flea (*Xenopsylla cheopis*). The plague bacterium (*Yersinia pestis*) was brought to North America by ships from Asia bearing infected fleas and rats in the late 19th century. The fleas and the plague spread to wild rodent populations and moved eastward. The flea is found on rats, prairie dogs, rock squirrels and other rodents, and also on dogs, cats and rabbits. Dogs and cats can become infected through the bite of infected fleas or by eating a rodent, rabbit or other animal carrying infected fleas. These domestic animals can also carry the fleas back into the house. Although 10 to 15 human cases are reported each year in North America, transmission from a human to another human has not been reported since 1925.

When John and Linda first went to the hospital, the infectious disease specialist examined them in the emergency room. Once the specialist heard that they were from New Mexico and finding the fever and enlarged lymph nodes, plague was number one on his list. He immediately put them into isolation and consulted with the NYC health officials and the New Mexico Department of Health. New Mexico state veterinarian Dr. Paul Ettestad corroborated that the couple lived in a plague area and they might have been exposed to infected fleas before traveling to NYC. He recommended doctors try to remove fluid from a swollen lymph node for testing. Dr. Ettestad also sent investigators to the couple's property in New Mexico to set rodent traps around their home and along a nearby hiking trail where nests and burrows of packrats (also known as woodrats) were abundant. Fleas were harvested from the trapped rodents.

Linda's diagnosis was confirmed as bubonic plague, which is an infection in the lymph nodes and the most common form of the plague. She was successfully treated with antibiotics and released from the hospital after 10 days.

John started out with bubonic plague, but it then developed into septicemic plague – an infection of the blood stream that quickly spread throughout his body. He developed serious complications, needing dialysis because his kidneys shut down and a respirator because of lung failure. He developed gangrene in his feet because of poor circulation, necessitating amputation of his legs between his knees and ankles.

Why was John's plague disease so much worse than Linda's? Was his diabetes a factor? Did he have more bacteria? Was the infection more advanced before he received treatment? These questions could not be answered.

Paradoxically, John and Linda had been quite aware they lived in a plague zone even before they came down with plague. The previous July, John discovered a dead packrat on their property and made sure it was tested by health officials. After they learned the rat and its fleas were infected with plague bacteria, they were even more careful to keep their dogs away from rodents, and they cleared the brush away from potential nesting areas and removed any outdoor food sources near their home. They routinely used insect repellent on themselves and flea control on their dogs. Even with these precautions, they had been bitten by infected fleas. As John's tests indicated, he was infected with the same strain of plague as was found in the fleas collected on their property the previous July and in November.

John was hospitalized for 100 days and finally made it home after months of rehabilitation. Meanwhile, the New Mexico Health Department put up signs to warn hikers about plague at trailheads near their property and their close neighbours were warned about the risk for infection in the area.

In retrospect, the caregivers at Karen's daycare were right to be concerned about fleas. It's true there was little risk of direct spread of disease to other children and cat fleas don't transmit serious disease. However, as people and their pets encroach upon wildlife habitats, the risk of picking up fleas that do carry diseases increases. When cats and dogs are allowed to roam, they might come into contact with rodents and other wild animals and bring these fleas into the home. In addition, climate change may expand the potential habitats of both wildlife and their fleas. Thus, people should take precautions to prevent flea bites when walking and hiking outdoors. And mothers (and fathers) should be concerned about fleas!

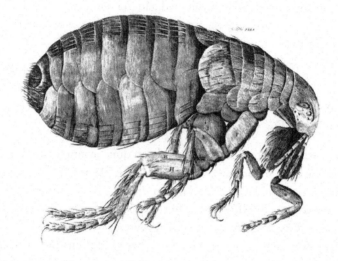

Figure 2 Hooke's drawing of a flea in Micrographia

"The strength and beauty of this small creature, had it no other relation at all to man, would deserve a description."
Robert Hooke in Micrographia (1665)

A Veterinarian Discovers a New Disease

Tracey McNamara

Dead crows had started to turn up in New York City around the grounds of the Queens Zoo, which was part of the larger Bronx Zoo/Wildlife Conservation Society, where I was head of the pathology department.

I was shocked to see so many dead crows because it takes something pretty potent to kill a crow. They eat just about anything, including road kill, so they are well adapted and tough. There had never been a recorded crow die-off in the area. It was clear that the death of this many crows in New York was a concern.

It was August 9th, 1999. Unbeknownst to me, this was three days after a peculiar disease had showed up in a human in the city.

Concerned that the wild crows might introduce disease to our captive collection, I immediately sent some of the dead crows to New York State's Department of Environmental Conservation, as they had jurisdiction over free-ranging wildlife. When I called a couple of weeks later to find out what had happened, I was dismayed to discover that no one had done any serious testing on the birds. The state's pathologist had simply cut open the birds' bodies and examined them superficially. I was told that they had found signs of metabolic bone disease, which wouldn't explain this spate of crows with neurologic signs followed by death.

As the weeks passed, we found more and more dead crows on the zoo's grounds. Then, some of the pheasants at the zoo began to display neurological problems. I determined the best course was to launch my own investigation by examining brains, hearts and other organs of the dead birds. I made microscopic slides of tissue, and performed bacterial cultures and stains and toxicology tests. I encased some tissue in paraffin blocks, immersed some in formalin and froze other tissue samples. It was essential to consider all possibilities.

To readers without a scientific bent, engaging in these steps may seem complex and time consuming, but I believe in traditional pathology practices. Besides, being a glass-blower and adherent of

Buddhist meditation, I have developed patience to do things methodically. I look at every tissue every time. Every creature that dies in our facility receives a complete post-mortem. To me, this is the culture of veterinary pathology, that of the older style generalist rather than a specialist, and it is crucial that you expect the unexpected as routine. It also doesn't hurt to have a basic level of humility, so you are willing to challenge your own diagnoses, along with the mental discipline to not jump to conclusions. Through this deliberate diligence, I was able to identify faint brain lesions and concluded that the pheasants had died from a viral infection of their brains (encephalitis).

On September 3rd, a little less than a month after the first dead crows were seen, I heard reports that St. Louis encephalitis virus (a virus carried by the culex mosquito, found mostly in the United States) was infecting people in New York. Encephalitic people? Encephalitic crows? I wondered if there could be a connection. That weekend I pored over textbooks and took notes to see if there was a connection between the human problem and the dead birds. The books confirmed what I already knew: St. Louis encephalitis does not affect birds. St. Louis encephalitis is caused by a flavivirus, which has never been a veterinary problem. Even with all this information in front of me, I knew that there still had to be a connection. We had encephalitis in humans and encephalitis in birds, so I had to at least consider the possibility.

Had I been taken seriously when I first sent the crow specimens and raised concerns, the mystery could have been solved weeks earlier. Instead, I was dismissed as some flighty veterinarian.

I had come by my love of my profession honestly and at an early age. I have a fond, enduring memory of visiting the NYC Central Park Zoo with my parents in the 1950's. In those days, the director of the zoo reared a black panther a la Kipling and even walked the cat on a leash through the park. Later, when I walked up to his cage for a visit, the panther was pacing back and forth. I said "Bagheera?" and the big cat stopped in his tracks, came up to the front of the cage and looked at me with his jade green eyes. That was it. I had chosen my life's work at age three. In time, I went to Kenya to volunteer with a team of wildlife researchers, who at first rejected me due to my lack of experience, but I would not give up and they finally relented. It was an invigorating experience that only reinforced my interests.

By the time I was probing this puzzling case in 1999, I was 45 and had developed a career as a veterinary pathologist specializing in wildlife diseases. Veterinarians are usually seen in a James Herriot mold of taking care of all creatures great and small. That's just not the case because the reality is more like George Orwell's *Animal Farm*: all animals are created equal, but some are more equal than others. For centuries, the emphasis has been on pets and on the economically important domestic species. That explains why today in the curricula of veterinary schools, there is substantially more attention given dogs and cats than whales and bats. With notions like that, perhaps I am a little different than the average veterinarian.

On September 7th, a flamingo and cormorant died at the zoo and even more birds were sick. The zookeepers at the Bronx Zoo loved these birds and were disturbed they were in danger of dying. I immediately performed post-mortems, first with the flamingo. I found massive myocarditis, an inflammation of the heart muscle, and in the cormorant there was extensive hemorrhaging in the cerebellum part of the brain. These were the most frightening lesions I had seen in 16 years as a comparative pathologist. It was frightening to realize we were dealing with something novel. One certainty was that whatever was killing the birds was lethally infectious and I immediately put on protection with a mask, goggles, coveralls and gloves to avoid exposure.

I had to intensify my post-mortems. I took impression smears and three sets of tissue samples from every organ to send to different laboratories. I also prepared 40 tissue slides for my technician. The next day, as we surveyed slices of tissue six microns thick, I witnessed the worst meningoencephalitis (brain inflammation) that I have ever seen. At the same time, I was painfully aware that I had more dead birds in the cooler, including a pheasant with massive cardiac (heart) necrosis and two flamingos with gastrointestinal lesions. I was desperate. It was unheard of for birds from completely different lineages and backgrounds to be taken down by the same disease. While this mystery was unfolding, 24 birds at the zoo, from a northern bald eagle to a black-crowned night heron, died or were so sick they had to be euthanized. Their clinical signs were shocking: birds staggering like they were drunk, laughing gulls dropping their heads, and a cormorant swimming deliriously in circles.

As the pressure mounted, I decided to approach the problem like a detective, revisiting every clue and analyzing the significance and interconnection of every element. I considered several possible causes: avian influenza, Newcastle's disease, or Eastern equine encephalitis (EEE). All were reasonable considerations, but they just didn't fit the circumstances. We knew these diseases targeted certain species of birds and those species of birds were doing just fine at the zoo. Had we been dealing with Newcastle's or influenza, the chickens in the petting zoo should have been dying from these well known poultry diseases. But the chickens were fine. Had it been EEE, the emus in our collection should have been sick as they are sensitive to that virus. But we had a healthy emu flock. Deductive reasoning pointed to this being an entirely new virus.

On September 9th, I phoned the Division for Vector Borne Infectious Diseases of the Centers for Disease Control's (CDC) in Ft. Collins, Colorado. One of the veterinarians from my team had stuck himself with a needle while euthanizing a flamingo. Convinced that the human and bird encephalitis disease were connected, I was concerned for his health. I pointed out to the CDC the unique coincidence of the St. Louis encephalitis outbreak in humans and the disease killing the crows and the zoo birds. I argued that more study was necessary. I asked them to test the serum sample from the veterinarian who had pricked himself with the needle and from the dead flamingo, but a CDC biosafety officer rejected my inquiry outright, partly because animals were not CDC's business or area of expertise.

The officer cut me off and pronounced that, "Birds don't *die* of St. Louis encephalitis. Birds are a *reservoir* for St. Louis encephalitis. And you should know better. You're just dealing with some veterinary thing." I was stunned by the lack of interest in scientific inquiry. I sent the serum to them anyway.

I also called the National Veterinary Services Laboratory (NVSL) in Ames, Iowa. Since they dealt with animals, I asked if they would test my tissue specimens for encephalitis viruses. I was relieved when they said yes. I knew they would test for all known encephalitis viruses, but I told them I thought we were dealing with something new and begged them to put my tissues onto cell culture as soon as they arrived. I was so convinced that we were dealing with a flavivirus, I told them I would eat my veterinary license if I turned out to be wrong! If the virus grew in the tissue culture and they got an isolate, they could use electron microscopy to confirm the presence of a flavivirus from the flamingo's brain based on the size of the viral particles. Fortunately, they listened. They called to say the specimens were negative for Eastern equine and other alphaviruses, but they had a virus they couldn't identify. When NVSL called me again a few days later, I was in the middle of another post-mortem. The told me the mysterious virus had started growing in the tissue culture and it was about the right size for a flavivirus like St. Louis encephalitis.

Eureka! Birds could die of a flavivirus, even if the textbooks said otherwise. The human and bird cases could be linked. This was exciting. Unfortunately, they lacked the technology to pursue the virus further. No flavivirus had ever been identified as the cause of animal disease so they did not have the reagents to further characterize the virus. This was something new to veterinary medicine.

I was pleased there was a link between the avian cases I was witnessing and the human problem and it served as validation of my diagnostic skills. Everything I'd been trained to do was paying off. Soon, however, the elation turned to anxiety. I had been cutting through tissues loaded with an unknown flavivirus and had definitely been exposed. I went home and wrote my will.

I immediately contacted the CDC lab at Ft. Collins and explained that a flavivirus had been discovered in the zoo's birds. By now, three veterinarians had pricked themselves with needles, so it was urgent to find out what might have infected them. Many birds were dying and the situation continued to be desperate. But the CDC doubted the results of NVSL saying they "were not certain of the quality of work done in veterinary labs".

Finally, it appeared we were nearing a breakthrough. The CDC learned through official channels that NVSL had isolated a flavivirus from the zoo's birds. Meantime earlier that week, the Connecticut Agricultural Experiment Station had announced the discovery of flavivirus from a crow and from mosquitoes, which supported my suspicion that mosquitoes were biting humans and birds. What remained to determine was the identity of the virus. Later that week, researchers at the US Army Medical Research Institute of Infectious Disease, or

USAMRIID, to whom I had sent samples as a last resort, had also isolated a flavivirus and were honing in on a diagnosis.

On September 20, NVSL sent my zoo specimen to the reluctant CDC to be tested and two days later the mystery was solved: it was a flavivirus - West Nile virus.

Interestingly, CDC officials phoned me soon after and without telling me they had found West Nile virus, they requested more tissue for testing. I was happy to oblige, though miffed at the delay in their interest.

Armed with my knowledge, I called CDC later and playing dumb, asked how their work was going. Of course, my specimens had unlocked the mystery. I happened to be in time to log into a conference call regarding the virus and when CDC reported its discovery, there was a pause. I jumped in. "Ladies and gentlemen, my name is Dr. McNamara. I'm a pathologist at the Bronx Zoo and these are my cases we're discussing. Would anyone care to hear the facts?"

Fortunately, I had kept thorough notes throughout the ordeal and I read through them. My review included whom I called, what I was told, and where and when I sent specimens. There was stunned silence when I finished. I had been pointing to the answer the entire time, but no one had listened.

It struck me that the problems with this crisis came down to a few personalities. It is worrisome that unnecessary obstacles can inhibit an investigation. At the time, there was little to no communication between veterinarians and the people who worked in the public health arena. In addition there may even have been some misogyny involved. I and some other women had raised the early questions, and it seemed to matter that the message was not being delivered by a male voice. The next time I call, I will speak in a baritone.

In the end, about 10,000 crows and 17,000 birds in total died in New York City alone.

Ten years after the West Nile virus outbreak, we still lack an integrated human/veterinary zoonotic disease surveillance network in the USA, but at least there is increased awareness of the contributions of veterinary medicine to public health. In 1999, it took someone who was able to look at disease across species to crack the mystery of West Nile virus. It took a veterinarian.

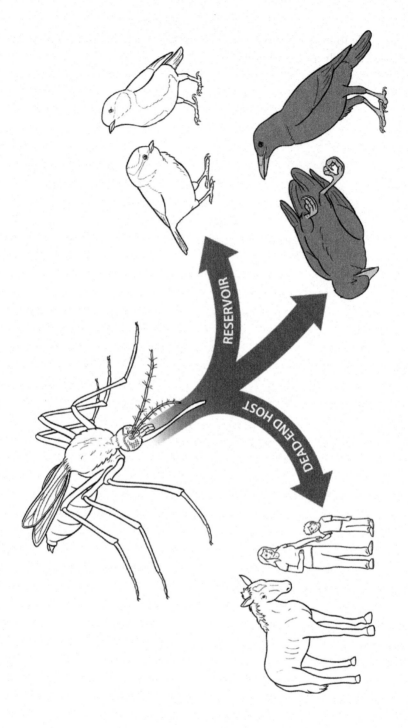

Figure 3 West Nile Virus

An Important Balance:
Keeping Dog-Patient Visits Healthy

Sandra Lefebvre

Bob's eyes glisten as he is hoisted up onto the bleach-scented sheets of the bed. Curly tail wagging, the toy poodle is quick to lick the proffered cheek of an elderly palliative care patient who beams with equal enthusiasm.

"My goodness," the old man cries out. "Who do we have here?"

"This is Bob," says Jana, the handler. "Say hello, Bob!" and Bob yips on command. "Bob sure seems to like you, Mr. Mutschler."

"Well, I just happen to have a cookie here for my little visitor." The translucent skin covering Mr. Mutschler's arm jiggles as the fragile limb stretches toward the food tray at the other side of the bed. Bob's tongue hangs from smiling lips, slick with anticipation.

The attending nurse stares incredulously, as Jana explains, "Mr. Mutschler doesn't talk to anyone but that dog!"

***

Every week Bob typically visits about 25 patients at a long-term care facility and countless others in a hospital in southern Ontario. Jana and Bob exemplify the perfect visitation dog team – spreading joy to patients, visitors, and staff - and in return, they enjoy the satisfaction of bringing comfort to others.

I've met many visitation dogs – from burly Newfoundlands to delicate Papillions. All of them had an ability to tolerate rough handling, a knack for ignoring startling sounds and strange stimuli, and a gift for making even the most grumpy or lonely people feel loved. Their handlers were equally remarkable, many devoting several days a week to visiting people who might benefit from interaction with animals. All handlers believed in the healing power of the human-animal bond and in the good that they were doing. For this reason, they agreed to participate in my study on visitation dogs.

I had already surveyed Ontario healthcare facilities and found great variability in policies and protocols for animal visitation programs. Some facilities prevented all animal visits because of concern about disease transmission, while others allowed animal visits but their guidelines were often incomplete and misguided. This was disconcerting. Unlike human visitors, animal visitors can get up on beds and lick patients and visit several patients, often in different wards. And of course, animals don't wash their hands – or paws! Thus, our research group wanted to find out what might be living on healthy-looking dogs in visitation programs.

During 2004, I traveled across southern Ontario, testing active visitation dogs for bacteria and parasites. We were particularly interested in 17 different zoonotic organisms – those that can cause disease in both animals and people. Even though all visitation dogs must be healthy, some positive findings were to be expected because bacteria such as *Pasteurella* spp are part of a dog's normal flora and some intestinal parasites may not be detected by the dog's owner, yet their eggs or larvae could be shed in the dog's feces.

We did find evidence of at least one zoonotic organism in 90 of the 102 healthy dogs we tested. Unfortunately, Bob, the poodle, was one of the dogs that tested positive for *Clostridium difficile* in his feces. As it turns out, Bob was the only dog in our study that was shedding a particularly virulent strain of *Clostridium difficile*, known as NAP1.

The NAP1 strain of *C. difficile* had caused outbreaks of severe infectious diarrhea in hospitals around the world, but had never before been reported in dogs. And no one knew if *C. difficile* could pass from dogs to people or from people to dogs. We felt it was important to report our finding right away, so we wrote a letter about our findings to the editor of the journal, *Emerging Infectious Diseases*. Much to our surprise and dismay, our letter sparked a firestorm of media attention - reporters started calling me about the letter even before it appeared in print. I was not prepared for this onslaught, having had no previous media experience or training. The headlines screamed that canine visitation programs were dangerous! Looking back, I realize that I had revealed more information than was prudent or responsible. Over the next month, I heard from angry study participants who felt I had betrayed their trust, and rightly so. I personally felt an albatross being draped around my neck, one that would weigh me down for several years.

I had already started recruiting participants for a new study on zoonotic bacteria that are important in the healthcare setting, including *C. difficile*. Our question was: When dogs started working in a visitation program, were they more likely to shed pathogens in their feces after they began visiting patients than dogs that were recruited into other types of animal-assistance activities, such as reading programs for children? In other words, if dogs didn't have the bacteria in their feces before they started visiting health-care facilities, did they acquire them during their visits? I recruited 96 dogs that visited healthcare facilities and 98 other dogs that did not. Their owners collected and submitted fecal specimens and nasal swabs every two months for one year.

Some of the owners were eager to participate because they were already concerned that their dogs might become ill from visiting hospitals. And this concern was valid – like humans, dogs can develop diarrhea from *C. difficile*. I shared their worry that these canine Samaritans could catch disease from people they visited. I had received an email from a woman in the USA whose mother had been hospitalized for *C. difficile*-associated diarrhea. Subsequently, the mother's dog also developed severe diarrhea with the same characteristic odour as her mother's. Testing of the dog's feces revealed the NAP1 strain of *C. difficile*. Unfortunately, we couldn't make a definitive statement about the relationship between the mother and the dog because the specific strain of *C. difficile* from the mother was not determined.

During the next year, my friendship with study participants blossomed. They graciously invited us into their homes and even invited me to potluck dinners and barbecues. I became attached to several of the dogs I met on a regular basis and developed a deep respect for their handlers.

As test results came in, I noticed a trend: dogs that visited healthcare facilities had *C. difficile* in their feces more often than those doing other work. In fact, I found that dogs that had visited health-care facilities were about twice as likely to have *C. difficile* in their feces as dogs that had not visited. Also, visitation dogs that had been treated with antibiotics were twice as likely to have *C. difficile* in their feces as visitation dogs that had not been treated with antibiotics. The dogs didn't seem to become ill from their visits and diarrhea was not noted in those dogs with *C. difficile*. We concluded that the dogs had acquired *C. difficile* during their visits to the healthcare facility.

To understand why, I followed some of the dogs as they made their visit rounds to see when and where they might pick up the bacteria. We wouldn't know if the dogs got the bacteria from a specific patient because the healthcare facility itself could have been contaminated with *C. difficile* spores. But we wanted answers to many questions - Did the dogs get the bacteria on their paws and then lick their paws? Did they lick the hands of healthcare workers who were contaminated? Did healthcare workers apply hand sanitizer before handling the dogs? Did the dogs go up on the patients' beds or visit patients in isolation or intensive care units? I also tested the dogs' paws and fur for *C. difficile* and other organisms before and after their visits.

I witnessed so many wonderful interactions between the dogs and people that I came to believe even more strongly in the power of the human-animal bond. Elderly residents in long-term care facilities delighted in the visits, some of which triggered memories of their own pets or confessions that their own families didn't visit them nearly as often as the dogs. Children in oncology units suddenly brightened; bored, bed-ridden patients in surgical wards took interest. People in waiting rooms went from anxious or hurting to delight when introduced to a dog. Even healthcare workers would stop what they were doing to pet the dog and temporarily relieve the stress of their work.

Some observations were not so heart-warming, and in fact, were alarming. I followed one visitation team through several wards – and then saw the dog drinking from a patient's toilet. At other times, doctors and nurses handled dogs, then cared for patients without washing their hands. Dogs sometimes licked patients on their mouths or healthcare workers on their hands. Several dogs repeatedly jumped from the floor to the bed and back again. Of course, anything they picked up from the floor might be transferred to the bedding.

As suspected, we found that climbing on patients' beds increased the risk that the dog would pick up *C. difficile*. These findings were worrisome – dogs with *C. difficile* from healthcare facilities might spread the bacteria to other facilities they visited, their homes and the community. With many dogs being coprophagic (eating feces), if a dog sheds *C. difficile* in a dog park, other dogs could become colonized or temporary carriers of the bacteria. We also learned that one of the dogs acquired *C. difficile* on its paws during the visit and that it was the same strain as the virulent one detected in the feces of the poodle, Bob. The troubling web of potential disease seemed wide and endless – and part of me wished we'd never initiated this line of inquiry.

After all, I was drawn to research on visitation dogs in the first place in part because I had visited hospitals with my cat as part of a programs run by a humane society when I was a student. Although I might have been curious about the possibility of disease transmission, I was much more interested in the beneficial effects of such programs. My positive impression from those early experiences was reinforced many fold as I witnessed dedicated dog-person teams work their magic.

However as my research wound down, I became disenchanted with the nature of my work and particularly averse to the word "potential." Potential does not mean inevitable, only possible. Even if dogs did pose a risk of disease transmission, weren't the many well-documented benefits worth that risk? And if proper infection control practices were used, couldn't we minimize the risk of disease? But just what *was* proper in such situations? Dogs could not sanitize their paws and the idea of sanitizing their fur was ludicrous.

With the albatross still swinging around my neck, I swore the next time my research group reported any findings and dealt with the media, we would be careful not to give out any sound bites that could be taken out of context. We also needed to develop guidelines for healthcare facilities so they could allow visits for dogs and cats even though we didn't have all of the answers. We had learned a lot from our research and the guidelines would help bring perspective to the intensified debate about the potential dangers of animal visitation programs.

In discussions with various stakeholders, we considered how we could prevent animals from picking up pathogens such as *C. difficile* from the hospital environment and subsequently spreading them. Some suggested dogs wear booties to keep their paws from being contaminated, but their fur or the booties themselves would still touch the bedding. Hand sanitizer had similar drawbacks.

Then why not stop the dog from going on beds? But many patients can't leave their beds, and these patients were likely to benefit from visits. Clearly another approach was necessary – one that would allow dogs onto beds with physician permission, yet protect both patients and visitation animals. As a result, we agreed that a disposable or cleanable barrier, such as a puppy training pad or pillowcase, should be used between the animal and the bed. Dozens of similar issues were tackled and strategies developed to maximize the benefits of patient-animal interaction without creating undo risk of infection. This thoughtful, collaborative undertaking led to the consensus guidelines that were published in 2008 in the American Journal of Infection Control. In response, many pet therapy organizations adapted and improved their canine visitation protocols accordingly.

***

Three years later, I found myself in the same hospital that Bob the poodle and I had visited. But this time I was in the surgical ward where my own 98-year-old grandmother was slowly succumbing to a twisted bowel. I was sitting by her bed with the rest of the family, awaiting her passing. I noticed a visitation dog – a Yorkshire terrier in resplendent dress - prancing down the hall. I jumped at the chance to talk with the handler, who was applying an alcohol-based sanitizer to her hands and had a diaper bag full of puppy training pads slung over her shoulder.

"Hi there," I said as I left the room to introduce myself. "I'm so pleased to see they allow visitation dogs in here. They never used to."

"They just started allowing us in a few months ago," said Fran, the proud owner of 'Mitsu.'

"We didn't have a protocol until recently," explained the charge nurse, who was escorting the visitation pair on their rounds. "We need to be particularly careful with patients in this ward because their immune systems are highly vulnerable."

The albatross's grip on my neck began to lighten.

"Have you had a chance to visit my grandmother?" I ask Fran, gesturing toward my grandmother's room.

"Oh, Florence," she smiles. "Yes. She really likes Mitsu, but seemed a little confused. Kept saying, 'Pinch me, boy!'"

I can't suppress my smile. "Ahhh . . . you must mean Princey boy!" I explain. "Prince was her dog many, many years ago."

And then tears finally escape from me in hot torrents that I do not try to dry.

"Are you okay?" Fran asks, placing her hand gently on my arm.

"Perfect," I reply. And at that exact moment, the albatross takes flight.

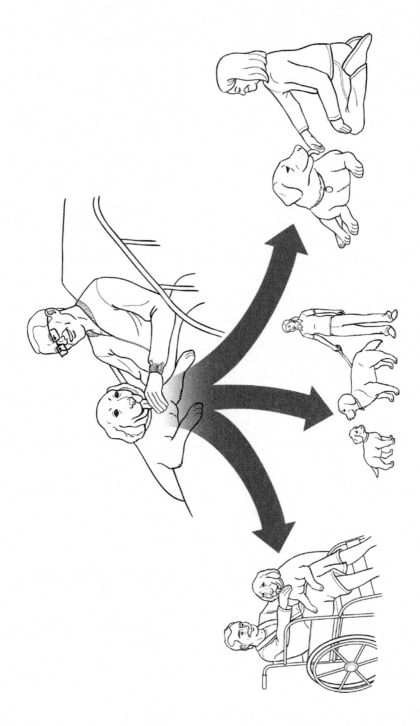

Figure 4 Pet Visitation Programs

Owls Under Siege

Bruce Hunter

Something was terribly wrong.

On July 20, 2002, The Owl Foundation in Vineland, Ontario was rocked with the unexpected death of Sultan, an 18-year resident great grey owl and one of 251 in the sanctuary. A day later a second great grey owl died, then a snowy owl. Despite all attempts at emergency care and treatment, birds became sick and died every day for several weeks. By September 15, 95 owls, 37% of the resident population, were dead. For founder, Katherine 'Kay' McKeever, and the staff, it was the loss of 95 friends and the impending loss of the dream.

***

Earlier that summer in May and June, The Owl Foundation was already reeling from a devastating loss. Larry McKeever, devoted life partner to Kay McKeever, had died at the age of 95. Moments of optimism had provided some healing: Young owls that had hatched during the previous year's breeding season were now almost a year old and the staff was looking forward to releasing these young birds to the wild. At the same time, the new hatch of 2002 was looking promising. It was a time of hope and expectation.

The Owl Foundation is the premier owl rehabilitation centre in North America and is the most successful facility in the world for breeding owls in captivity. Every owl at the Foundation represents a small miracle.

All the resident birds were once free and wild somewhere in North America, but had been found injured, ill or orphaned by a Good Samaritan with enough interest and compassion to seek medical help. With the help of the wildlife rehabilitation network, these owls from across Canada and parts of the USA had eventually found their way to The Owl Foundation.

On arrival at the Foundation, an owl is carefully assessed and treated. Those birds with injuries too severe to allow them to be released to the wild, yet with the potential to become breeding owls, are given new homes at the Foundation. Although their injuries range from blindness to loss of hearing to missing limbs, each owl retains its ancestors' genes. The work and dream of The Owl Foundation is to ensure that this genetic diversity is preserved through the release of the offspring of these owls.

Successful breeding of owls is an art and a science. After years of studying the behaviour of owls and a background in architecture and design, Kay has designed elaborate, expansive spaces that satisfy the breeding needs of the resident owls. Quiet and seclusion are essential for successful bonding and breeding of owl pairs; thus, monitoring of each bird is through an elaborate video system. Kay and her team know all the owls intimately, having devoted countless hours to discretely observing and recording their behaviour. Each owl is an individual with its own quirks and personality and each one plays an important role in the Foundation's work.

The Owl Foundation specializes in breeding species of owls from the northern boreal forest and the Arctic - an amazing breeding collection that included majestic great grey owls, semi-diurnal Northern hawk owls, boreal owls, diminutive saw-whet owls and arctic snowy owls.

Before the disaster struck, The Owl Foundation housed breeding pairs of all 16 species found in the various regions of Canada.

**

Autopsies on the 95 dead owls carried out at the Ontario Veterinary College at the University of Guelph ruled out all common diseases of owls. With the emerging West Nile virus as a suspect, samples were sent to the Canadian Science Centre for Human and Animal Health in Winnipeg, Manitoba – Canada's only level IV biocontainment microbiology laboratory. By mid-August, West Nile virus was confirmed and this West Nile virus outbreak became the largest die-off ever recorded in a captive bird collection.

**

The West Nile virus story in the western world began in New York in 1999, where the virus was linked to the deaths of crows at the Bronx Zoo and two human deaths. The virus survived the winter in mosquitos and in 2000 began its rapid spread across North America.

Within three years, the disease had been recorded in most USA states and Canadian provinces, Mexico and the Caribbean. It is now firmly established in the wild bird population and the North American strain is closely related to genetic strains in Israel and the Middle East. No one knows for certain how the virus arrived in North America, but scientists speculate it may have been carried across the ocean in an infected bird or perhaps by mosquitos in the luggage of travelers.

West Nile is a member of the *Flaviviridae* family of RNA viruses and is related to other important human pathogens, including dengue viruses, yellow fever viruses and certain encephalitis viruses found in North America, such as St. Louis Encephalitis virus. Since first identified in Uganda in 1937, the West Nile virus has been mainly associated with mild infections in people in Africa, the Middle East and Europe, causing fevers, headaches and muscular weakness. Less than one per cent of infected people develop the neuro-invasive form, with neurologic signs such as tremors, cognitive impairment and occasionally severe encephalitis and meningitis (inflammation of the brain and spinal cord, respectively). About one-half of those affected with the neuro-invasive form develop longer-term neurological problems.

Although we generally think of West Nile virus as causing disease in people, it is primarily a virus transmitted from bird to bird by mosquitoes. Once the bird is infected, the virus replicates in the tissues and during certain periods of infection, it enters the bloodstream. A mosquito feeding on a viremic bird (viremia = virus is in blood) acquires the virus with the blood and passes the virus on to another bird when it takes its next blood meal. While some species of mosquitoes feed entirely on birds, others species are more cosmopolitan and will also dine on people, horses and other mammals. These mosquitoes are called "bridge vectors" and are the critical link in the chain of infection by the West Nile virus. They are responsible for West Nile virus transmission from birds to people and horses, and occasionally other mammals such as dogs, cats, skunks, small rodents, and bats.

For a person to become infected, there must be viremic birds, the proper species of mosquito that feeds on both birds and mammals and good environmental conditions, such as warm temperatures and standing water to ensure enough mosquito activity for virus transmission. Mammals are generally considered to be "dead end hosts" of the virus because even though they become ill and could die from the infection, they rarely develop high enough levels of virus in their bloodstream to allow for transmission of the virus by mosquitoes to other hosts. Person-to-person transmission is rare, although there have been a few cases in a patient who has received an organ transplant or a transfusion of a large volume of blood from an infected donor.

To control the West Nile virus, mosquito populations must be reduced by eliminating pools of standing water where mosquitos breed and by spraying chemical insecticides. In addition, people should use personal mosquito protection such as repellents and long clothing, and put mosquito screening on their houses, and stay inside during periods of peak mosquito activity.

***

The West Nile virus devastation at The Owl Foundation brought despair and hopelessness to the staff and at times they felt an overwhelming urge to abandon their dream. But tragedy often brings out the best in people. Kay and her staff had incredible strength and resilience and a growing determination that even in death, these owls had a purpose that could further the dream of conservation. Many questions remained unanswered so The Owl Foundation and scientists at the Ontario Veterinary College agreed to use this disaster to learn more about this newly emerging disease.

The research team included scientists from the Ontario Veterinary College, others at the University of Guelph, the Canadian Cooperative Wildlife Health Centre, and Health Canada in addition to The Owl Foundation. Careful autopsies on every dead owl revealed that different species of owls had very different susceptibility to infection with the virus. Owls that bred in the boreal forest and the Arctic, including great grey owls, Northern hawk owls, boreal owls, Northern saw-whet owls and snowy owls, were highly susceptible. These birds were clinically ill for a short period and then died with a death rate of 100%. Microscopic examination of their tissues showed minimal inflammation, which indicated that they died so rapidly from infection their bodies had almost no time to react.

Species with a wider breeding range extending from the northern boreal forest into the USA, such as the great horned owl, the barred owl and the short-eared owl, had intermediate susceptibility to the virus. Only occasionally did a bird in this group become ill and the illnesses were longer and had a lower death rate. In the birds that did die, post-mortem examination demonstrated that their body defences had reacted significantly to the virus before they died. Of the owls of species that breed further south, such as the Eastern screech owl, the common barn owl, the flammulated owl, the burrowing owl or the long-eared owl, none became ill or died from West Nile virus even though they were housed in areas where birds of other species became ill and died.

This remarkable difference in the susceptibility of different owl species to West Nile virus has been shown in other bird species. The American crow and other members in the family corvidae (such as ravens, blue jays and grey jays) are extremely susceptible to West Nile virus infection. In fact, crows serve as the sentinel species for surveillance programs for West Nile virus because they die quickly from the virus and with their large size and distinctive appearance, they are easily recognized by the public. Furthermore, crows inhabit populated areas in almost all of North America, providing an early warning that there is virus activity in a region. In contrast, some species, such as the American robin, the house sparrow and other small passerine species, are important species that maintain and transmit the virus. They rarely become sick or die from infection with the virus, but they have a long period of viremia when the virus is accessible to mosquitos.

Researchers studying the owls that died at The Owl Foundation also learned that West Nile virus could be transmitted from one bird to another by other biting insects, such as flat flies (also known as hippoboscids). When the researchers microdissected flat flies found on dead owls, they found that the stomachs of the flat flies contained West Nile virus. It is now recognized that the virus can be transmitted by many blood-feeding insects, including flies, mosquitoes, ticks and others.

Researchers also evaluated whether or not a vaccine developed to protect horses from West Nile virus would be effective in owls. Some species of owls appeared to develop a response after being vaccinated, however, others did not react at all. Thus, vaccination, at least with the technology of the time, was not considered to be a reliable way to protect birds.

Armed with the results of these studies and others, The Owl Foundation spent thousands of dollars to mosquito-proof all outdoor cages of the northern forest and arctic species of owls that are most susceptible to West Nile virus. In addition, a special mosquito-proof isolation facility was built where susceptible birds could be housed during peak periods of mosquito activity in the summer. An enhanced control program for flat flies and also an enhanced disease surveillance program were implemented.

Now, years after the devastating outbreak of West Nile virus, injured and crippled Northern owls still arrive at the Foundation and new breeding pairs of owls continue to form and have offspring carrying their genetic makeup and being released to the wild. Data on owl behaviour are still being collected, helping to inform new, improved cage designs. The pain of 2002 will never be forgotten, but life at the Foundation has returned to an even keel.

Epilogue

West Nile virus is an excellent example of how an existing pathogen, well-established in one geographic area, can cause significant and widespread effects when it moves into a new region, undergoes small mutations in its genome and is introduced to animals or birds that have never encountered the agent before. West Nile virus is the most important disease to affect wild bird populations in North America in the last century. The effects of this zoonotic virus have been dramatic and have affected wildlife agencies, zoological facilities, infectious disease researchers, ornithologists, population ecologists, public health agencies, government policy makers and the general public. The spread of this virus has had significant effects on local populations of birds and has caused widespread concern and panic in the public. Government and local health agency responses in developing disease surveillance programs and providing healthcare and educational information have been extensive and expensive. West Nile virus has raised awareness of vector-borne diseases (spread by insects and other arthropods) and is a perfect example of the interconnection and complex relationships between animals, people and the environment.

Ruminations on the Consequences of a Life

Brian Evans

Prologue: The Seeds of a Cowtastrophy

It was the Friday before the long May weekend and Sam was in great spirits. The holiday weekend traffic had not yet started so he was making good time completing his deliveries. As he put his turn signal on and began to slow his rig to make the turn into the laneway for his final stop of the day, he looked down at the dashboard clock and smiled. He would be back to the feed mill by 2:30. That meant if he allowed himself an hour to clean out the truck and complete his paperwork, he could swing by the butcher shop to pick up some steaks for the barbecue, hit the liquor store for supplies and also stop at the flower shop and still be home in time to surprise his wife before she got home from work.

Humming along to the Jimmy Buffet song playing on the radio, Sam glanced to the pasture on his left where he saw a number of midnight black cattle gazing in the direction of the truck and the trail of coffee coloured dust that rose behind it, even at his reduced speed. It was almost as if they were saluting his arrival.

Then his eyes locked on a heifer calf standing beside her dam. Something about the calf caused Sam to stop the truck while he was still 100 yards short of the gate to the farmyard. As if orchestrated by some unknown and unseen conductor, more and more heads turned to acknowledge the now stationary truck.

Sam realized he had stopped humming and reached down to turn off the radio. He was fixated on the young heifer. The silence that now enveloped him added to the reverence he was experiencing. The feeling of some warm saliva dribbling out of the corner of his mouth, down his chin and onto his chest made him realize his mouth was open and caused him to look away. Glancing down briefly, he saw a brown stain blossoming beside the company logo on his blue shirt. He really should break his habit of chewing tobacco.

Again his eyes drifted back to the pasture and the several hundred cows with their recently born calves in close proximity. As if drawn by a magnet, his visual field narrowed like a zoom lens until it focused directly on the heifer calf, standing regally and proudly, that had first caught his attention.

The blaring of a truck horn from the farmyard ahead of him served to break the spell. As Sam looked forward once again he saw Francis, the ranch foreman, waving at him from behind the gate with the big, bright crimson STOP AND REPORT FOR BIO-SECURITY sign. Seeing Sam's wave of acknowledgement, Francis undid the chain and rolled back the gate to allow entry.

Engaging the gears and easing back on the clutch, Sam drove into the yard and, with Francis serving as his flagman and ground crew, he was guided back to the auger to unload the feed from his truck into the designated holding bins.

Jumping down from the cab, Sam smiled sheepishly at Francis. "Sorry for the pit stop in the lane. You must have thought I had broken down. I was just checking out the herd on my way in and, for some reason I can't explain, one of the heifer calves just caught my attention. There is just something very special about her."

Francis grinned as he removed his faded baseball cap with the remnants of the Hockey Canada logo and dragged his calloused hand over his head and neck to wipe away the sweat. "No need to explain to me, Sam. Her name is Molly and damned if you aren't the fourth person this week who has commented on that one. I've been telling the boss the same thing since the day she hit the ground."

A half hour later, with the feed off-loaded and a slap on Francis's shoulder, Sam hoisted himself into the cab of his truck and drove out of the yard, through the gate and down the lane. As he approached the highway, he came to a stop and could not help but glance over his right shoulder for one last look. There stood Molly, like some black bovine beacon in a sea of green. Shaking his head, Sam turned on the radio and the strains of Meatloaf filled the air. Before he resumed his ritualistic humming, Sam stuck a fresh plug of smokeless tobacco behind his lower lip as he muttered to himself, "Molly, you are indeed destined to be remembered."

Molly's Cowtemplations

The early morning sounds of the farm served to rouse Molly to consciousness. As she slowly opened her eyes, she noted that the day was dawning to reveal the potential for a cloudless sky. The sun was barely starting its long, slow, daily climb over the eastern horizon, but already the narrow shafts of light framed a sky that seemed to run on forever. "Big Sky" country was how the locals fondly referred to it.

Even though the sun lacked any demonstrable warmth so early on this January morning, it seemed to offer hope of better days to come. Surely it was a good day to be alive. But then again being alive was always better than the alternative, right? Given the way Molly was feeling, it was certainly better to focus on the positive. The last few weeks had been difficult for her both mentally and physically.

With more effort than should be required, or that she could ever remember it taking before, she slowly turned her head. From her position in the pen, she could see several of her herd mates in the adjacent field rising and stretching in anticipation of the new day. Easy for them to do, she thought. When was the last time she had felt well enough to even make the effort? It must have been at least three or four days ago.

The growling sound of a tractor starter turning over in the distance followed by the throaty roar of the engine as it sprung to life served as a catalyst to the last members of the herd to get to their feet. That sound was a prelude to the imminent delivery of several round bales of some of the finest alfalfa hay you could imagine. Funny, since she had been a calf at her mother's side, she was always the first in line at the trough, a dominant force in the social hierarchy of not only her birth herd but in each successive herd she had been in over the course of her life. Somehow the thought of being number one did not hold much attraction for her this morning. Even the prospect of eating today was not high on her list of things to do.

Casting her big brown eyes about the pen, she began to feel more despondent and alone. The "sick pen," they called it. An appropriate name she thought. Equally applicable as an acronym . . . Separated, Isolated, Controlled, Kennel. Although the thick bed of straw on which she lay was comfortable to a degree, her hind legs felt like they were in never-never land.

Recognizing that her mood was quickly becoming as black as her coat colouring, Molly sighed and then watched her warm breath turn into a white mist as it hit the cold morning air and then started to dissipate as it drifted skyward. If only the fog that had increasingly been lingering in her head at the margins of her consciousness would also lift and dissipate, perhaps she could make some sense of what was happening to her.

Forcing herself to focus on better times, her mind wandered back through thoughts and memories that seemed to surface in a series of random and disjointed sequences. Come on brain, get it together, she urged. Closing her eyes, an image gradually formed and emerged from the recesses of her thoughts.

Another location, another time. She was in a warm, comfortable, secure place. It was dark. It was quiet. It was the only place that she had ever known. Then, inexplicably, she felt the walls that had defined her existence begin to move. Gradual and gentle at first, over time the movements became more rhythmic, more intense. The contraction of the walls seemed to take on a sense of purpose and urgency. Against her will, Molly felt herself beginning to move. She was powerless to resist. Suddenly she saw light as she emerged from her safe place and slid out onto the bed of what she would later come to know as straw. Compulsively she took a deep breath and air filled her lungs for the first time. She was cold, she was wet, and she was confused. However, before she could panic, she felt the warm, loving and reassuring caress of her mother's tongue. Slowly, methodically, the licking continued from the tip of her nose to the top of her head and over her entire body. Yes, it felt good to be born. It felt good to be alive. The circle of life was intact. Look out world. Molly has arrived!

From those first moments, Molly had always had the heartfelt conviction that she was special and had a purpose to be fulfilled. She was the offspring of a planned mating. On her maternal side was a long line of excellent brood cows known for their ease of calving and strong mothering abilities. On her paternal side, the production traits ran deep and true with good birth weights and healthy, vibrant and fast maturing calves. With genes like hers, there had never been any doubt in her mind she was a blessed and gifted cow. But it was more than just her bloodlines and her physical attributes. She was born with a destiny and she had this innate sense that somehow she would have a profound and lasting impact.

The thought of her pedigree and her place as the premier cow in the herd served to raise her spirits and resolve. Giving her head a shake to clear the fog that seemed to be threatening to roll in and shroud her thinking once again, she allowed the memory to fade. In its place, she was forced to confront her current situation once again.

When had things started to change? Was it before or after the first snowfall? In retrospect, the more she thought about it, the changes probably started almost three months ago. At first, the signs had been subtle and only noticeable to her. That is when she first had the sense that "something" was different. But she couldn't define what that "something" was. She couldn't quite put her hoof on it, as they say. Funny how humans had a similar expression, but they referenced "finger" instead of "hoof". "I wonder whether cows came up with the expression first or if humans did?" Molly thought to herself. It really didn't matter in the big scheme of things, but it was exactly these types of rambling, incoherent thoughts crossing her mind that she now recognized were likely the earliest onset of the changes.

The mental wanderings were then followed by a slight change in her emotional well being that was not observed by any of the other members of the herd. For the first time in her life, her sense of personal self worth, invincibility and self-confidence began to develop cracks. She had an underlying level of anxiety that came and went. It caused her to be more sensitive and react impulsively to events going on around her, such as sudden sounds and movements.

She had written it off as possibly a hormone thing. Having been a mother herself now five times, wasn't she entitled to mood swings? Or maybe it was just the time of year. After all, every autumn the rancher would take stock of the herd and make decisions as to which animals would stay and which would go. Her pedigree, sense of destiny and status in the herd notwithstanding, her next birthday would be number seven. Her qualities as a premier brood cow were above challenge. She had already been sold twice before. In both cases, she was sure it was because she was something special. What rancher wouldn't want her as part of the foundation of his breeding program?

However, with continued improvements in the herd breeding programs, it was expected her offspring and their subsequent generations should be genetically equal or superior to her and the generation before. The three successive daughters she had birthed on the two previous farms where she had lived were a part of her living legacy. The same was true for the daughter she had delivered following her arrival on this farm two years ago.

Last spring Molly had given birth to her first male offspring, while her most recent daughter had given birth for the first time. A granddaughter! "A truly mooooving experience! Udderly fantastic and behooving someone of her status." All the cow accolades poured in.

Shortly after all the calves had been weaned, Molly's son and all the other male calves had been sold. That had been the first time Molly had experienced seeing one of her own being moved out. Perhaps that explained, in part, why she felt somewhat emotionally unbalanced.

She had taken great pride in the fact that her granddaughter had followed in her daughter's hoof prints (Didn't humans have a similar saying about that too? Something to do with footsteps? Darn, those mental mind wanderings were back!) and had been selected to be a 4-H calf by the rancher's daughter. She was not surprised when her granddaughter had been selected best in show. After all, she had Molly's genes and it further reinforced Molly's belief that she really was something special and had a destiny to make some profound and lasting impact.

But autumn had come and gone. The culling decisions were made and Molly was still here along with her daughter and granddaughter . . . but still, the mental wanderings and occasional bouts of anxiety persisted. In truth, rather than diminishing they came more often and lasted for longer periods.

It was at that point that she had begun to hear the whispers of some of the other cows. Between chewing their cud and belching, cows have too much time on their hooves (hands?) with nothing better to do than offer their opinion on the state of the world or the latest rumour-du-jour about their herd mates. Standing in the fields all day, there is a lot of gossip that goes on.

In Molly's case, there seemed to be a preoccupation by the herd with the fact that she was mellowing, perhaps losing her edge as the lead cow, and whether her daughter was ready to assume the mantle. It was not a democratic process and frankly none of their business. They talked about how she was occasionally distracted and, on more than one occasion, could be seen standing apart from the herd.

One day after she appeared to trip crossing the field there was a chorus of accusations that maybe she was spending a little too much time licking the salt and mineral block. Molly was indignant. She never over indulged. However, she had to admit that walking a straight line could no longer be taken for granted and getting up and down was becoming a challenge.

The past month had been a difficult time. The physical manifestations had become more pronounced. The occasional slight stumble gave way to a wobbly gait. Her hind end felt weak and almost as if it was disconnected from her body. It seemed to have a mind of its own while Molly was convinced she was losing hers.

Then, four days ago, she could not get it together at all. She could not get her legs to respond to her mental commands. After delivering the hay bales, the rancher had spent 20 minutes urging her to rise and join the others. Between his gentle pushing and her full concentration, she had finally gathered all her strength and will and made it to her feet where she stood, not sure how to command herself forward. Patiently, the rancher slowly guided her in her unsteady state to the pen that had since become her fortress of solitude.

The deep guttural roar of the tractor penetrated her thoughts. Looking up, she saw two large round bales had been placed in the centre of the pasture and broken open like a bovine culinary banquet. In moments, the herd had come to the table set for them and began to feast.

Rather than heading back to the farmyard to continue his daily routine of chores, the rancher brought the tractor up to the side of the pen. It was then that Molly noticed the large flat sled called a stone boat being towed behind. The rancher opened the gate to the pen and positioned the stone boat beside her. Then he slipped large canvas belts under her chest, one behind her forelegs and another just ahead of her hindquarters.

Taking his seat on the tractor once again, the rancher honked the horn three times. In response, in the distance she heard another motor starting which, in the ensuing minutes, grew louder and louder until a large enclosed truck appeared. The driver manoeuvred the truck until the open back end was adjacent to the sick pen gate.

The next thing she knew, the driver of the truck had raised the two canvas belts and looped them over the forks on the front of the tractor. He gave a two-thumbs-up to the rancher ("Easy for him to do, impossible for me," Molly thought to herself. "I don't even have thumbs"). Ever so slowly, the forks of the tractor inched skyward until the straps took up the slack. Molly felt them tighten around her as she was being gently lifted.

When she was at standing height, the truck driver pointed to his left. Sensing rather than knowing the intent, Molly called up every fibre of her being and instructed her legs to walk onto the stone boat. Relieving the weight of having been lying on her right hind leg for the last two days felt wonderful. It was almost as if the buzzing and tingling in her right hind leg was a sign of the circulation returning blood supply to her muscles and the nerves reconnecting to her brain.

Then she felt the tension go out of the sling around her chest as the rancher lowered the forks and she was placed gently down on the stone boat that had been covered with a thick blanket of hay in a sort of ruminant room service. The straps were removed and the rancher repositioned the forks under the stone boat and began to lift it.

As the platform was raised, it was almost as if she was being lifted up on an altar or a throne. How fitting! Her regal manner returned as she looked out over the rest of the herd. They all stopped eating and fixed their eyes on her. She was above them all. She truly was special! Seeking out her daughter, their eyes met. Afterwards they said she nodded her head ever so slightly and in that moment, the baton was passed. Her legacy was intact.

The rancher loaded the platform with Molly into the back of the truck and the truck began its slow procession through the farmyard.

He had a few words with driver about ensuring a safe and smooth ride to the small abattoir some 30 kilometres away where he had made arrangements to have the carcass prepared for his freezer.

Then the rancher stepped back, looked up at Molly and saluted, looking her straight in the eye and said, "You were, are and will always be one unforgettable cow."

Then he shook his head and walked away, muttering to himself, "I have no idea why I just did that."

On arrival at the abattoir, Molly was inspected by the veterinarian, who noted her medical history, and the signs of central nervous system problems and the muscle and lung damage caused by her lack of movement for several days. Thus after she was killed, the veterinarian decided to condemn the carcass and prevent its use for food.

In addition, because of her condition and her history, he sent samples of the brain to the laboratory to check for bovine spongiform encephalopathy (BSE), more commonly known as mad cow disease, a disease that had never before been confirmed in a Canadian-born animal. As a precaution, the Canadian Food Inspection Agency had established an ongoing surveillance program to detect any possible cases.

Epilogue: The Cowsequences

In May 2003, BSE was confirmed in a cow in Canada by provincial, federal and international reference laboratory experts. BSE is a progressive neurological disorder of cattle caused by an unusual transmissible agent called a prion. Although not well understood, a prion is thought to be a modified form of a normal protein known as prion protein that has morphed into a pathogenic (harmful) form that damages the central nervous system. A cow can contract BSE by ingesting food that contains nervous system tissue with prions during the cow's first year of life.

In humans, most cases of transmissible spongiform encephalopathy are believed to occur spontaneously. However, since 1996 a variant form has been identified that is associated with the possible consumption of infected tissues. There is no cure.

With the diagnosis of BSE confirmed in the cow, the largest Canadian animal disease investigation in several generations was initiated. It had three fundamental objectives: 1) locate the farm of the birth herd of the positive cow; 2) determine how the feed that was consumed during the first year of the cow's life could have been contaminated with BSE infectivity; and 3) identify, trace and test all other cows born on the same farm within 12 months of the positive cow that may have eaten feed from the same source.

To reconstruct the movement of this cow over the previous seven years, investigators had to review records and sales receipts as well as interview ranchers, truckers and livestock sales personnel who may have had knowledge of the cow's history. They ended up with a veritable spider's web of interactions, assembly points and movement patterns that identified more than 80 possible farms and approximately 2,700 associated animals of interest in four provinces.

Investigators worked with forensic experts and used blood typing of animals and review of breeding records to identify genetic relationships between possible dams or sires from the various farms. It took more than five months before an eventual match to the birth herd was confirmed. Testing of the several thousand animals of interest identified over the course of the investigation did not reveal any further cases of BSE.

The confirmation of BSE in a domestically born and raised animal in Canada had profound economic impacts on the cattle industry, estimated in the billions of dollars, and also on animal health and public health policies in the country. These included an increased awareness and emphasis on the importance of animal identification and traceability, a prohibition on the transport and slaughter of cows that aren't mobile, and the mandatory removal of tissues that may harbour BSE infectivity from all animals slaughtered in Canada for human food.

If Molly was that cow, then there can be no doubt that she fulfilled her destiny and made an irrevocable difference. Fortunately, since BSE is not a disease that can be passed on from mother to offspring, the legacy of her progeny remains intact.

Authors' disclaimer: *This story is a work of fiction. Any similarity between any two-legged or four-legged characters in this story and any persons or animals living or dead is entirely coincidental. Because it's fiction, be assured no cows were harmed in the writing of this story.*

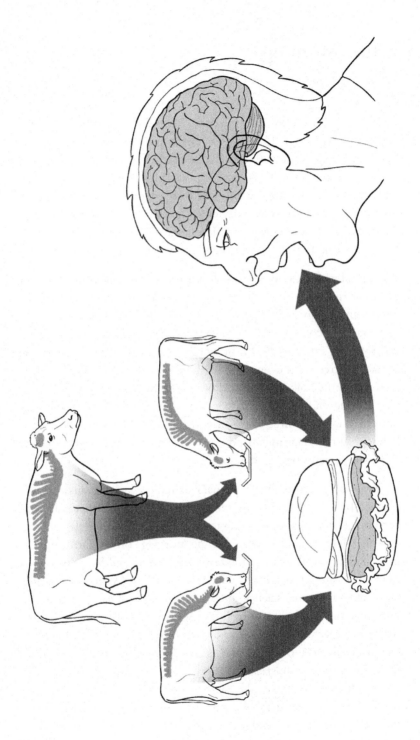

Figure 5 Mad Cow Disease

A Bad Morning

Paula Menzies

Helen lay in bed staring into the dark. The clock radio read 5:30 am.
Vern, her husband of 15 years, was in the kitchen making coffee
before heading to the barn to milk the goats. The smell drifting up the
stairs nauseated her. She could not rise. It had been another sleepless
night marked by sweats, dizziness and a hard, painful cough. The
doctor's appointment was today at 9:30, but it felt like a decade
away. She had a growing conviction it wasn't flu that had gripped her
this past week, but how could she make her doctor understand that?

Several weeks earlier, the smell of fresh coffee had enticed her out
of her warm bed and into her morning routine, leaving the children,
Emma and Danny still asleep. The farmhouse was old and the
upstairs floor cold. Fortunately, winter's back had been broken with
last Saturday's thaw. On this first day of spring, pools of melting snow
lay in the driveway – even at this early hour - and mud was thick in the
yard where the milk truck turned during its twice-weekly pick-
up. Throughout the winter, the volume of milk in the bulk tank had
been down, but in three weeks, the main doe herd was due to start
kidding. When their milk production peaked a month later, so should
Vern and Helen's bank account.

Two years before, they had finally exited the swine industry –
plagued by low prices and mysterious new pig diseases that made their
accountant nervous while trying to balance the books. Helen's
parents, Heinrich and Freida, held the mortgage on the farm and lived
in a bungalow at the end of the farm lane. Helen and Vern had met at
the University of Guelph and with their new BSc (Agriculture) degrees
and majors in animal science, they had married and taken over her
family's farm. Managing a small farrow-to-finish pig operation, they
were progressive farmers and leaders in the community at adopting
new technologies. But the swine industry's troubled times started to
wear down their enthusiasm and they began looking for another
commodity where they could use their skills and entrepreneurial
spirit. A trip to a farm show piqued their interest in the dairy goat
industry. There was no expensive quota to buy as with dairy cows.
And there was a growing consumer demand for goat's milk, driven by

a wealthy middle class looking for new and interesting foods and an expanding ethnic population more accustomed to milk from goats than from cows.

In the fall of 2008, Helen and Vern purchased a herd of Saanen goats from a retiring farmer that included 50 bred does, due to kid in the spring, and seven bucks. They also purchased a group of 50 bred does at a local farmer's production sale that were due to kid within a few weeks.

Goats normally breed only in the fall and give birth in the spring. Milk production by the doe peaks about four to six weeks after she gives birth and drops off totally after nine to 10 months. This means that goat dairies normally produce the most milk throughout the summer. Then, the herd is "dry" or not milking through the late winter months when the goats rest before their next kidding. But cheese makers who produce the fragrant herb-encrusted chevrai – soft, crumbly cheeses that melt in the mouth – want to make cheese in the winter as well as summer. Thus farmers get a premium for milk produced through the winter months. Fortunately, the Loewen's 50 doe herd gave birth uneventfully in October, 2008 and the winter milk sales had provided the Loewens with much needed cash.

The family fell in love with their goats. The Saanens were big and white with mischievous, intelligent expressions. Their unending curiosity could be frustrating because anything new was investigated and challenged before being accepted. They were also personable. They would readily come to the side of the pen and nibble on coats and mittens – begging for scratches behind the ears

Now on this spring morning in 2009, Helen was on her way to feed the goat kids from the 150 does that had been born earlier that spring. The kids were bottle-fed colostrum (liquid rich in immune factors) that had been carefully heated to kill viruses and bacteria that might affect the goats' health. This program had been setup by their veterinarian, Dr. Mackenzie. Emma and Danny fed and cared for the adorable and clever kids, which quickly learned the two children were the source of milk and grain. The goat kids would crowd the children as they brought the milk into the pen, leaping, rearing on their hind legs and chewing on pigtails and hat strings, much to the delight of five-year-old Danny, who liked to wrestle with the older bucklings.

Emma, 12, who was thinking of becoming a veterinarian like Dr. Mackenzie, would fuss over the kids' health. Diarrhea was viewed as a disaster and necessitated immediate treatment with an oral electrolyte solution to keep the kid hydrated until the disease ran its course. When Dr. Mackenzie visited, Emma would follow her and offer important history on animal well being, particularly of the kids. On this morning, Helen was feeding the kids to give Emma a chance to sleep off a minor cold. After taking care of the kids, Helen's next stop was a pen of pregnant does due to start giving birth in a few weeks.

These doelings were offspring of the original herd of 50 does and the first generation to be raised on their farm. The family was proud of their growth and stature and were looking forward to the litres of creamy white milk they would supply. When Helen entered the pen, she noticed one of the doelings was hanging back from the feeder and looking about anxiously. She walked back to examine it and saw two tiny kids – clearly premature – lying dead on the straw bedding. The doeling was trailing a piece of afterbirth (placenta), which Helen took away so that it wouldn't make the animal sick. Helen had seen spontaneous abortions before and the occasional premature loss was normal on any farm. She wasn't too worried and placed the fetuses and placenta in an old feed bag for disposal on the manure pile. The doeling had little milk in its udder. Helen sighed, thinking that if the doeling can't be milked, she'll have to be shipped to slaughter, not a good start to the much anticipated kidding.

The next afternoon, Emma raced into the house very upset.

"Mom, there are dead kids in the pregnant doeling pen!" she exclaimed.

Helen had not mentioned the first abortions to Emma to not upset her, but she had not expected more. She and Emma went to the barn and sure enough two more doelings had aborted. They were both at the feeder enjoying their hay, but the tiny fetuses lay in the pen, lifeless. Vern was milking and the noise of the equipment made talking difficult, so she waited until he came to the house.

"I think we have a problem," she whispered after taking Vern aside. They had seen similar losses in the sow herd years ago and knew that abortion diseases could affect goats, but their does appeared so healthy they thought it couldn't happen on their farm.

In the morning, they called Dr. Mackenzie, who arranged for the fetuses and placentas to be sent to the Animal Health Laboratory at the University of Guelph for a diagnostic work-up. Later that day, the veterinarian visited the farm and examined the doelings from the affected pen. Another one was aborting. She took a careful history and raised an eyebrow when Vern told her about the 50 does purchased the previous fall at the farm sale.

"Did any of them abort?" she asked.

Vern looked embarrassed. "Yes, two of them did abort but I buried the fetuses. Both does were pretty close to term and they seemed fine and they are still in the milking string," Vern said, glancing at Helen. "Sorry, I thought it was OK. Just trying to save money on veterinary expenses."

Dr. Mackenzie recognized the tension. The couple was clearly concerned that they had brought something horrible into their herd.

"The main point is that we are investigating things now," the veterinarian said, then settled on a straw bale and took the couple through the plan she had developed. "Unfortunately, it is too late to stop your doelings from aborting this time. By the time the first doe has problems, the damage to the fetuses and placenta in the other animals is usually well advanced."

Helen felt like crying.

Dr. Mackenzie continued, "Infectious abortion is a common problem in goats and has 3 main three causes – toxoplasma, chlamydia, and coxiella. Toxoplasma eggs are only shed by felines, both domestic and wild; I know you don't have any cats in your barn so I am most concerned about chlamydia and coxiella. The diagnostic tests will tell us which one is the cause. With chlamydia (*Chlamydophila abortus*), we can give a vaccine to the rest of the goats to protect them in future pregnancies. But if it is *Coxiella burnetii*, better known as Q-Fever, there is no vaccine available in Canada. Even more important, I'm worried about the risk to your own health."

Vern and Helen regularly read the farm papers but had only vaguely remembered Q-Fever.

"What is Q-Fever?" Vern asked.

"It's caused by bacteria that can infect any animal," Dr. Mackenzie explained. "Not just goats but also sheep, cattle, cats, dogs, rodents, pigs, wildlife, birds, arthropods – and most importantly, humans. Infected animals shed it in their manure, birth fluids and milk. When an infected goat aborts, there is literally a cloud of organisms around it. The bacteria get into the environment, they are very hardy, and can live for months in a spore-like state. When manure is spread on the fields on a windy day, the spores can be blown up to five kilometres away."

"If it can infect any animal, why do goats and sheep get the bum rap?" Vern ventured, remembering an article he had read in a newsletter.

"Goats, sheep and cats appear to be most often affected by the disease," the veterinarian replied. "When a pregnant goat gets infected for the first time, she often aborts just like your doelings. Usually about 15-20% abort, but it can be as high as 40%, so it can be pretty devastating. Other animals like dairy cattle aren't as badly affected and rarely abort, but they may shed the bacteria in their milk. The main problem with Q-Fever though is that the disease is highly contagious to humans."

"What do you mean?" said Helen, her maternal instincts ignited. "Could Emma and Danny get sick from the goats?"

"Anyone is at risk, but the most susceptible are pregnant women, very young children and those with poor immune systems such as the elderly or people ill with another disease," Dr. Mackenzie explained.

Helen thought of her parents living only a few hundred metres away from the barn. Her father had recently gone on heart medication. Her anxiety level was rising.

"But not everyone who gets infected becomes ill," the veterinarian added, trying to be reassuring. "60% don't even realize they have an infection; 20% may feel like they are getting the flu, but recover in a couple of days, never becoming ill enough to go to the doctor. But 20% get severely ill and need treatment to recover."

"So, it's treatable?" Vern asked.

"Oh, yes, very treatable with the right antibiotic. The important thing is if any of you become ill in the next few months, go immediately to your family physician and ask to be tested for Q-Fever. You have to be your own advocate. Prompt treatment is important because once the disease becomes chronic, it's much harder to treat."

This was scary news.

"And you need to protect yourself," she continued. "Wear a special mask when you work with your does during kidding; the masks should filter out 95% of dust and organisms in the air. Also, wear long-sleeved disposable gloves during kidding or if you handle the placenta. Wash your hands frequently with a good antibacterial soap. And make sure your barn clothes stay in the barn. That includes coveralls, boots, coats, mittens and hats."

Helen looked over at Vern's head and gave a small smile. He was wearing his much beloved feed store hat (only thing they ever gave him for free, he joked). It went with him everywhere, even occasionally to the dinner table. Now it was going to live in the barn for sure.

"Absolutely we can do that," Vern said. "But what about the goats? Can we treat them, vaccinate them?"

"Antibiotics might slow the number of abortions," the veterinarian explained, "but there's debate on how useful they are. Usually the abortion storm runs its course and the herd develops a level of immunity so that abortions become sporadic. But the goats continue to shed the organism in their birth fluids, manure and milk, sometimes for weeks or months after kidding. "

Helen was getting frustrated, asking, "So what do we do now?"

"Now, we wait for the diagnosis from the lab," Dr. Mackenzie said. "We can treat the doelings with antibiotics, but we have to withhold their milk until it is clear of the antibiotics. After that, the milk is fine for sale as long as it is pasteurized. In the meantime, I suggest that you keep the children and visitors out of the barn, bury or burn any more aborted fetuses, and monitor your own health."

The bad news from the laboratory came at week's end: Q-Fever. Five more doelings had aborted but none in the last 24 hours. By now, a quarter of the group had aborted and none were coming into milk. What a blow! Helen and Vern did the chores with little enthusiasm, unsure of what their future would hold. The children picked up on the tension. Emma could not understand why she couldn't go in the barn. Sick goats were her specialty, she so wanted to help her parents. The goats in the barn that had been such a joy a week ago now brought trepidation and dread.

Helen's father showed up at the barn for his milk. Vern and Helen knew the dangers of raw milk and the disease organisms that could be present – they pasteurized all the milk used in their home. But Heinrich avoided the pasteurizer and got his "quart" straight out of the bulk tank.

"Such nonsense," he would say to Helen. "I am 65 years old and I've been drinking raw milk since I was a baby."

"Yes, but think of the times you were ill and blamed it on mother's cooking," Helen retorted.

Heinrich stood at the barn door, perplexed by the hand-written sign that said "Dad, I love you but stay out of the barn." Helen had called him after the veterinarian's visit and told him of the risk to his health, but he still considered himself immortal. The sign had been put up after two polite and respectful verbal warnings. Helen had a jug of cold pasteurized milk ready for him. "Dad, this bug is in the air, in the manure and in the raw milk. I don't want you to get sick. Please." Heinrich saw the stubbornness in his daughter's eyes and left with the milk.

Helen and Vern had closely followed Dr. Mackenzie's biosecurity recommendations. It wasn't a big change as they had always been cautious when they had pigs. The difference was that instead of worrying about bringing a disease to their pigs, they now worried about bringing a disease into their home. The rest of the doelings started to kid and, apart from a higher than expected stillbirth rate and some weak kids, the doelings came into milk well and seemed fine. Maybe life would return to normal, just a different normal.

Then came the morning when the coffee made Helen nauseous and she could barely move. She had gone to bed early the night before, feeling like an actor in a pain relief ad. Fever, headaches and muscle aches. She was exhausted from the work and the worry and thought a good night's rest would fix things, but this morning she felt no better. Pop more pills and keep going. It was time to visit the doctor.

When she arrived at the office of Dr. Best, the family's long-time physician, she had a fever of 38.5° C, a cough that shook her whole body, and was dizzy with a headache. The doctor examined her, feeling lymph nodes and looking at her throat.

"You look tired," he observed.

"You noticed," she cracked back.

"It looks like the flu," the doctor said. "Here is a prescription for antibiotics. Take one pill a day for seven days. If you're not better by then, make another appointment."

After a pause, she asked: "Could this be Q-Fever?"

The doctor looked puzzled. He didn't want to appear ignorant in front of a patient but he couldn't quite remember ever hearing about the disease in medical school.

"No, just the good old flu and maybe a touch of walking pneumonia. That's what the antibiotic is for," he said, a bit paternalistically. "Where did this idea of Q-fever come from?", he thought. "Probably too much internet surfing."

Seven long days later, Helen finished the antibiotics, but felt even worse. Her bedroom looked like a hospital, with pill bottles, glasses of water, throat lozenges and tissue boxes. After the visit to the doctor, she had given up going to the barn. Not because of his advice, but because she was too weak and faint. Vern's concern for his wife was all over his face. He'd brought her food, but she was too nauseous. Vern sat beside her on the bed and handed her the pain pills. It was the new family ritual in the morning.

"I'm going with you to the doctor's this morning," Vern insisted. "You aren't getting any better and I am losing faith in his abilities."

"Dr. Mackenzie called yesterday," he went on. "She asked how the kidding was going and if everybody was OK. I told her you had the flu and were feeling pretty tough. She asked if it was Q-Fever and I said that Dr. Best had said no, that it was a bad case of the flu. That seemed to set her off a bit."

Dr. Mackenzie was nice but sometimes got a little perturbed when someone didn't take her seriously.

"She said that human doctors don't know much about zoonotic diseases, particularly Q-Fever," Vern reported, "and sometimes they don't take new information from their patients without a little prompting. She emailed me a fact sheet from the government. She suggested, strongly I might add, that we bring it with us when we visit Dr. Best today."

Helen didn't like challenging Dr. Best's diagnosis but took comfort that Vern would be there to fight her battle.

Dr. Best read the Q-Fever fact sheet that Vern had politely but firmly given to him. After a long minute, Dr. Best looked up at Vern.

"You have animals at home with this disease?" he asked.

"Yes, and I had more than a half dozen abort about three weeks ago."

"I must admit, I don't remember much about this disease from med school," Dr. Best added. "But so many diseases got the 'five-minute mention'. Helen's illness has gone on too long to be just a virus, but I still think it's good old walking pneumonia. It can knock the stuffing out of you and takes awhile to get over, but it should have responded to the antibiotic - and she's still got quite a fever."

Vern watched the doctor's face. Clearly he was considering this new possibility for a diagnosis.

"I'll have the nurse pull a blood test for Q-Fever," the doctor continued. "Also, let's get some chest X-rays at the hospital and see how bad those lungs are." Vern breathed out slowly. *Now we are getting somewhere.*

The chest radiographs were so disturbing that Dr. Best immediately admitted her to the hospital. He had seen lots of patients with pneumonia during his years in practice, but this pattern of infection was different. The radiographs showed circular dark spots in her lungs rather than the large wedge-shape typical of most pneumonias. Dr. Best called the local health unit. Fortunately, Dr. Berubé, the county medical health officer, was in his office.

"What do you know about Q-Fever? I think I've got a case here," said Dr. Best, and then shared the story of Helen's history, the goats and the abortions, as well as the details of her illness and failure to respond to antibiotics. "The X-rays show an unusual pattern, not typical of the pneumonias I usually see. Also the patient's liver enzymes are really high."

"Sounds likely that your patient has Q-Fever," Dr. Berubé responded. "While you wait for the Q-Fever test to come back, change the antibiotic treatment to doxycycline, and keep it up for two to three weeks. Also, keep her in hospital until her fever is down for 48 hours. If it is Q-Fever, you should see a response within a few days, but it may take more than a month for her to completely recover. She then needs to be monitored to make sure the Q-Fever doesn't become chronic."

Dr. Best scribbled the prescription on his pad while Dr. Berubé continued.

"How is the rest of the family?" he asked. "They should be checked as well. It's a very treatable disease if caught early, but dangerous if ignored."

Two weeks later, the whole family was in Dr. Best's office. Helen's test had come back positive and the health unit had recommended testing everybody who lived on the farm. Danny was pulling toys out of the box in the waiting room, looking for his favourite truck. Emma, now feeling much more adult since her mother's illness, was trying to keep his energy under control. Helen, still weak and coughing, put a gentle hand on her daughter's shoulder, indicating her gratitude.

Dr. Best looked at the lab results and said, "Vern got infected with the bug but must have been one of the lucky ones who doesn't get sick. The children and Mr. and Mrs. Schroeder didn't even get infected. Your veterinarian and her biosecurity precautions were on the ball for this one. Sorry I didn't listen to you when you first came in, Helen."

A few weeks later, Dr. Mackenzie dropped in for her regular herd health visit. There had definitely been an increase in abortions in the mature does that had just finished kidding, but it was only 10% and not the horrible 25% as in the doelings.

"It's in the herd and it will not be going away anytime soon, but you shouldn't see as many abortions in the future," she explained. "Your herd is going to be healthy again and will produce more milk than ever. But we have to keep up the same biosecurity precautions with your family."

"Forever?" Helen asked.

"Yes, at least until we get that European vaccine licensed here in Canada. It will reduce shedding if the herd has a solid vaccination program and, as far as I can determine from the experts, it is the only viable long-term option for the small ruminant farmers."

Several days later, Helen lay in bed with the morning light from the spring sun streaming across the floor. The children were sitting on the bed holding a tray shakily between them. On the tray was Mother's Day breakfast. Along with the slightly burned toast and overcooked egg, with juice and coffee slopping onto the tray, was a beautiful hand-made Mother's Day card with Emma's well-drawn goat kids jumping through flowers under a big, yellow sun. Danny's tractor appeared to be chasing them across the meadow. The sentiment was timeless – and Helen's heart was full. Vern came upstairs with a bouquet of flowers – daffodils from the garden. It was a good morning.

Antibiotics of Mice and Men

Claire Jardine

"Oh, what a mess," I thought as I surveyed the kitchen, with sticky, pink goo on the walls, the floor and all over the high chair.

"I hope some of that got into you," I said to my eight-month-old son, Peter, who was now laughing delightedly as the dogs licked him and his chair. "Oh well, at least the dogs are cleaning it up," I thought as I grabbed a washcloth to assist.

Peter had just finished his last dose of oral antibiotics that had been prescribed for a bacterial skin infection. The syringe that I had been given to measure and administer the medicine had become stuck, and then, just as I applied my super mom powers to the plunger, it came unstuck and sent the pink medicine flying into Peter's mouth and beyond.

"Oh crap," I thought as I looked at the clock "Ben," I yelled, "I've gotta run."

Ben, my husband and today's babysitter, came running down the stairs.

"Oh dear, what happened here?" he asked.

"The syringe jammed," I told him as I handed over the washcloth.

"Ok, I've got it," he said, and took over as I blew kisses and ran out the door.

As I biked to work I tried to make the shift from home life, where I look after my family and inexpertly squirt antibiotics all over the walls, to my work life where I am supposed to be developing my research program. I was a new mom and a new faculty member just back from maternity leave, which made for a busy, and at times, stressful life.

I grabbed a coffee on the way to my office and met one of my first graduate students, Sam, in the hallway. He looked really excited and handed me a paper.

"Hey Sarah," he said, "I just read this really cool paper on antimicrobial resistance in wildlife in the U.K. and I was thinking that maybe we should see what is going on here in Ontario."

"Great, I have a couple of meetings this morning, but stop by my office this afternoon so we can discuss the paper and your ideas," I said, dropping off his paper on my desk and heading off to my morning of meetings.

After lunch, I settled down to read the paper Sam had handed me. Researchers in the U.K. had found that some bacteria they had isolated from the feces of wild mammals were resistant to antibiotics. As far as anyone knew, these animals hadn't been given any antibiotics. Exposure to antibiotics is thought to be a key reason that bacteria in a person or animal might be resistant to antibiotics, so this was an interesting finding.

"Cool eh?" said Sam as he pointed to the paper on my desk.

"Definitely," I replied, "but I think I need another coffee before we get started."

"Are you buying?" asked Sam as we headed out the door.

As we settled into comfy chairs at the coffee shop across the street, I asked Sam why he was so interested in this.

"Well," he said, "I think it is really important. Resistance to antimicrobials is a big problem. My girlfriend had cystitis that was caused by bacteria that weren't affected by the antibiotics she was taking. She was really sick and it took quite a while for the doctors to figure out why she wasn't getting better. People infected with these resistant bacteria are more likely to have longer hospital stays and require treatment with antimicrobials that may be less effective, more toxic, and more expensive."

"Yes, that is important," I agreed, "but how does it relate to the paper you found about resistance in the bacteria from wildlife?"

"Well, the thing is," he said with a smile, "it's like you always tell me, it's all interconnected. We don't just worry about resistance in humans, we also worry about resistance in other animals. Not only can it make bacterial diseases in animals difficult to treat, which is a challenge for veterinarians and farmers, but also because these resistant bacteria can be transmitted to humans. If a bacterium like salmonella, for example, gets transmitted from a cow to a person and the person gets sick, that's bad, but we can treat it with antibiotics if necessary. Now imagine that the salmonella is resistant to the antibiotics that we usually use to treat the disease and you can see that situation is *really* bad. Also, what is even more frightening is that even if the resistant bacteria don't cause disease in humans, they can transfer their resistance to other bacteria that could go on to cause disease."

"Okay," I said, playing the skeptic, "I can see why this is a concern when we are talking about farm animals and our pets, but I need a bit more convincing before I get too worried about resistance in wildlife."

"Well," Sam said, "we know people can come into close contact with wildlife, think about mice droppings in your cupboards, and that wildlife can come into contact with livestock and crops on a farm, so there is a chance that resistant bacteria from wildlife could spread to people and other animals. We also know that some wild animals can move large distances. Those species could be spreading resistant bacteria throughout the environment."

"Plus," I exclaimed, "it is just so interesting! So, what exactly do you want to do?"

"Well, I thought we could look at what is out there in Ontario wildlife with an approach similar to what they did in the U.K. study," Sam said.

"That would be interesting," I said, "but perhaps we can do even more. If antibiotic use leads to the development of resistance, and we know that wildlife aren't deliberately treated with antibiotics, then we need to figure out if they are being inadvertently exposed to antibiotics. A lot of antimicrobials are naturally produced by other bacteria or fungi, so wildlife may be exposed to naturally occurring antibiotics in the environment.

"Wildlife also could be exposed to antimicrobials used for livestock or to livestock or human waste that is contaminated with antibiotics." Sam elaborated, "It would be cool to see if the levels and patterns of resistance in wildlife varies in different environments."

I agreed it was a good idea and left Sam with the task of developing a research proposal. It was a great relief to see him so excited. We had been tossing around a number of ideas for his graduate research, but this was the first time he had shown real enthusiasm for a project. Although my main research focus was on wildlife disease and not antimicrobial resistance, I knew there was a lot of research expertise in antimicrobial resistance at the University of Guelph, so while Sam was developing his proposal, I made connections to ensure that this project would be successful.

Early the following week, Sam emailed his proposal to me explaining how we would look at antimicrobial resistance in wildlife living in urban, rural and natural areas. He had decided to focus on wild small mammals that don't travel far, such as mice, voles and shrews. This made sense because it meant that what we found in the wildlife would reflect what they picked up in the local environment. Because I wasn't an expert in the genetics of antibiotic resistance, it would be important that Sam have other professors to talk with about his ideas and to help him with the lab work. I spoke to several researchers who were interested in Sam's proposal and keen to see if we would find resistant bacteria in wildlife in Ontario. I was relying on my limited funding to keep Sam going and to pay for his field work; luckily for me, they were interested enough to do the laboratory work for the cost of supplies.

The next couple of months were occupied with filling in permits and animal care approvals that are required to ensure that animals used in research are treated appropriately. We also had to identify and get permission to use all the sites we needed for trapping. In addition, we hired a summer student, Marie, to assist with trapping. Then, suddenly, it was summer and we were ready to start.

Unfortunately, the week we were scheduled to start, my husband Ben was away finishing his own field work and I was supposed to be on "holiday," taking care of Peter until my mom came the following week.

"Since we were planning to use my house as part of an urban residential area anyway, we might as well start there," I proposed, so we waited until Peter had fallen asleep to set the traps.

To trap mice we used live traps baited with sunflower seeds and stuffed with cotton so the mice could have a snack and bed down in comfort. That evening we set 100 traps in my neighbourhood beside buildings and composters, under junk piles, in gardens, along fallen trees, everywhere we thought mice might go.

Bright and early the next morning everyone met at my house to check the traps. Peter was already awake so I put him in his high chair where he was safely inside but still able to watch us working outside. We had just finished checking the traps when he began to cry.

"Oh darn," I thought and dashed inside to give him a cracker.

As Peter settled down to munch happily on his cracker, we settled down to work just outside. We had caught seven animals, which was not a bad start. I grabbed a trap and gently shook our first critter from the trap into a bucket.

"A deer mouse," I said as I grabbed it with a gloved hand. "See his bi-coloured tail and white belly, and look at those big brown eyes and big ears. So cute. We need to weigh him, record his body length and sex, tag him and let him go. The most important thing that we need from this little guy is the droppings from the trap, but we'll get those after we are done with him."

All went well with the processing and in just a few minutes the mouse was released back where he came from, hopefully no worse for wear. Next, I took the bedding out of the trap and, using clean forceps I picked droppings out of the bedding and put them into a sterile vial that we would take back to the lab. We labeled the tube, filled in the data sheets, carefully cleaned the forceps and then we were ready for the next mouse.

Inside the house, Peter was smiling happily as he munched his cracker and watched us with apparent fascination. "Yeah," I thought, "not even a year old and already starting field work, that's my boy."

"Okay Sam, you're up," I said.

I helped him with the next few mice and then it was Marie's turn and we were done processing by noon. Sam and Marie caught on quickly and I knew with a few more days of practice they would be pros. Sam took the samples back to the lab and we agreed to meet that evening to re-set the traps. Our plan was to set the traps for three consecutive nights at each site to try to maximize the number of animals we could catch at each site. I helped Sam and Marie for the rest of that week and part of the next and then they were on their own. They continued trapping throughout the summer and were able to meet our target of ten urban, ten natural, and ten rural areas by the beginning of September.

The lab work was very time consuming. For each of our samples the laboratory had to grow bacteria from the droppings and then test each for resistance to 15 commonly used antibiotics. Because we weren't paying clients we had to wait until the laboratory could fit our samples in, a small price to pay for getting the samples processed practically for free!

Early in October, Sam came running into my office, exclaiming, "Results, we finally have some results!"

"So what's the scoop?" I asked.

"Well, it's interesting," he explained. "We found resistant fecal bacteria in wildlife in all the environments, including the natural areas, but the prevalence of resistance to some antibiotics is quite a bit higher in wildlife living on farms with livestock than the other areas. I still need to do all the stats, but it is quite cool and I think our results show that livestock farms may be associated with a higher prevalence of some kinds of antimicrobial resistance in wild small mammals."

"Very interesting," I replied. "Okay, your next step is to go ahead with the statistics. Come back and see me when that is done and we'll chat more."

He stopped on the way out. "Hey, I forgot to tell you your backyard is a hotbed of antimicrobial resistance in wildlife. The prevalence of resistance was higher in your backyard than in any other urban residential area."

"Really?" I asked, surprised, "but lower than what you found on livestock farms, right?"

"Right," he laughed.

"Wow, that is crazy," I thought as I scooped my computer into my backpack. "I wonder what is going on in my backyard."

I jumped on my bike and as I headed home I thought about all the possible sources of resistance to antibiotics in my own backyard. If it was naturally occurring antimicrobials, why was it so much more prevalent in my backyard than in my neighbour's? As I biked down my street I noticed the remnants of a garbage bag that had been torn to bits, likely by crows, and the garbage was strewn all over and accessible to any creature. We put Peter's diapers into those same type of bags and when Peter was on antibiotics, there were likely resistant bacteria in the waste. I added garbage to my list of possible sources of resistant bacteria for wildlife.

I walked into the kitchen and was greeted by squeals of delight from Peter as two happy dogs cleaned up yogurt from Peter's hands and face. I couldn't help thinking back to that day last spring when I had squirted antibiotics all over the kitchen. The dogs helped clean the antibiotics with the same dedication they used to clean up food. So that is one way pets could be exposed to antibiotics. But more importantly, pets are also treated with antibiotics and when they deposit their waste in my backyard they could be contaminating the environment where birds, chipmunks and mice find their food. So I added pets to my list of possible sources of resistant bacteria.

As I cleaned the yogurt from Peter's chair, I thought about how the antibiotics on the cloth used to clean up the mess I had made in the spring were rinsed down the drain and, even more importantly, how antibiotics in human waste are flushed down the drain whenever someone is taking antibiotics. So water was added to the now fast growing list. And this was just my house!

It is easy to see how livestock farms might be potential sources of antibiotic resistance for wild animals, but it is also important to consider less obvious sources in our everyday lives that contribute to the development of antimicrobial resistance in wildlife. Not only are we contributing to the spread of resistant bacteria to wildlife, we are also contributing to the spread of resistant bacteria throughout the environment. This may not seem like a significant problem now, but we are inadvertently increasing potential sources of resistant genes that can be acquired by disease-causing bacteria. In addition, the impact of these resistant bacteria on naturally occurring bacteria and the environment is largely unknown. There is a lot to think about and lots of interesting research to be done.

I have always tried to be environmentally responsible, but the results of this work on antimicrobial resistance and wildlife have started me thinking about all the unintended impacts that I am having on the environment each day. Now, as I bike into work each day, I worry less about making the shift from home to work and realize that it really is all interconnected.

Memories of Giardia

Alastair Summerlee

Whenever it rains or the wind blows, I'm reminded of the Isle of Skye, a place of many childhood memories and the source of my passion for the environment, for biodiversity and for animals. But it is also likely the source of the zoonosis that I carried unknowingly for many years.

Every summer, our family would decamp to the Isle from our home in the south of England. Called the "misty isle" because it is covered, for most of the year in mist, drizzle and driving winds from the Atlantic Ocean, the Isle of Skye is part of the Inner Hebrides. The large island is dominated by the black Cuillin Hills (a range of mountains that was reputedly once the size of the Himalayas), the smaller Red Hills, and an extensive coastline of sea lochs, staggering cliffs and sandy beaches. The sea literally boils with fish of all kinds and there are dolphins, whales and sharks. The shoreline is covered with crustaceans of every variety and the cliffs are home to thousands of birds.

And the land? Well, the land is mostly covered by Scottish blackface sheep. In the 1700s, the Scottish landlords, or lairds, were persuaded by their English counterparts to engage in sheep farming. Fearful that local people would compete with the sheep for the land, they shipped whole villages of people from Skye (and many parts of the Highlands and Islands of Scotland) to new homes in Canada, the United States, Australia and New Zealand. The so-called "highland clearance" was one of the most remarkable uprootings of people in recorded history and is the reason why there are so many pockets of people around the globe claiming a Scottish heritage and bearing Scottish names.

Loch Eishort (pronounced 'Ish-short') was a magical place for children. We spent our summers dabbling in the rock pools around its shores, hunting for edible crab, scallop, mussels and oysters at low tide, searching through the seashells that accumulated on a special island that we called 'coral island', swimming in one lagoon or another, and fishing, walking or just tearing about the hills.

Directly in the path of the Gulf Stream, Loch Eishort not only benefits from warmer waters and climate, but the ocean current brings weird and wonderful creatures from the Caribbean. Enormous jellyfish with straggling long tentacles would pulse along below our boat in the crystal clear water, their eerie beauty belying the pain their armouries of stinging cells could deliver. Some days the water would be calmed by harmless moon jellyfish so great in number that it was hard to row the boat. Extraordinary velvet swimming crabs would emerge, as if from a magician's hat, as you pulled up the lobster pots and we could catch lythe, mackerel, spur dogs, skates and salmon to eat.

The beauty and wonder on the macro scale was mirrored by micro diversity. I delighted in stirring the water on short summer nights causing the phosphorescent zooplankton to generate a storm of light - a miniature, electric miracle of nature. Unfortunately, I had no idea that giardia was also ubiquitous.

Now to the less pleasant part of my story that, I should add, comes with warnings for those readers with sensitive dispositions. Do not read further if you have a tendency to weakness of the bowels. This story may be too close for comfort.

For much of my early life, I was prone to protracted and unexplained bouts of diarrhea. It was a marvellous way to balance a healthy appetite with a thin figure, but it did have its embarrassing moments and at times the abdominal cramps and gurgling became too much. Usually though, the condition was manageable. Stressful times in my life seemed to be associated with bouts of diarrhea, but I accepted it as a fact. But it became a serious problem once we had children of our own and I started to notice that my bouts of diarrhea coincided with diarrhea in my children. Weight loss in children can be dangerous, so all of us were checked out by doctors. We were diagnosed with giardia and we started treatment with metronidazole.

Known in Canada as the agent of "beaver fever," giardia infection results from ingesting cysts in contaminated water or food. The cysts can resist cold, heat, dehydration and survive for months in cold water such as the mountain streams fed by the peat bogs of Skye and even in city water reservoirs. Although there are distinct forms that seem to have a preference for different animal species, giardia from one animal species can infect another and vice versa. As most beavers carry the human form of giardia (*Giardia lamblia*), I suspect that the beaver is sadly maligned by the epithet, "beaver fever." It is likely that beavers in Canada were infected by early settlers and not the other way around.

Once ingested, enzymes in the intestine of the new host stimulate the cyst to hatch and a small hairy-looking trophozoite emerges and begins to feed actively. After splitting in two, the trophozoites latch onto the intestinal wall, aggravating the intestine and compromising its function. Most healthy people have either a short period of diarrhea and stomach cramps or may not exhibit any symptoms at all. Usually, the host rejects the parasite and it is shed within 24 hours, but in some people it stays.

After reviewing my past history, the general consensus was that I most likely had harboured giardia since my early days in Skye. But why had I continued to be victimized by the parasite? Did I have Crohn's disease, HIV or other nasties that weakened my immune system and allowed the giardia to stay? A battery of tests was run. While I was being tested, we were experiencing the horror of feeding metronidazole (a bad tasting drug) to three young children four times a day. (My son still has an abhorrence of swallowing pills that I blame on the treatment regimen). In the end, all the tests for other diseases were negative and I was in the 1% of the population who carry the protozoan for no known reason. I was started on a rigorous routine of treatment so I would shed my gut lining and remove the parasite once and for all.

Since that time, I have avoided red meat and dairy products on medical advice and I am pleased to say that my relationship with giardia, a zoonosis, is over. Of course I'm pleased to be rid of the bug, but now I have to watch my diet: it is easy to get fat when you are not losing weight every now and then. Still every time that it rains or the wind blows, I think of those microscopic hoards that once controlled my weight and I think of the "misty isle" and happy days of my childhood by the seashore.

Ebola and the Hot Zone:
A Veterinary Odyssey

Nancy Jaax and Jerry Jaax

The Class of 1973 for the College of Veterinary Medicine at Kansas State University College had an unusual feature - eight women among the 80 graduates. At that time the veterinary profession was dominated by men and veterinary school classes, certainly at Kansas State, typically had only one or two women. Nancy Dunn (now Jaax) was in this anomalous group. While in veterinary school, she met and married Jerry Jaax from the preceding year's class.

While Nancy Jaax finished her final year of veterinary school, Jerry accepted a direct commission to the Army Veterinary Corps. The Vietnam War was ongoing and the army was hiring (33 out of 80 from the class of '72 went into military service). The plan was for Jerry to serve his two-year commitment then the couple would open a small animal practice together. Instead, Nancy decided to also join the army after she graduated. The pair spent two years in the state of Washington and then took a three-year assignment to Germany, where they experienced a busy tour, including working with public health and zoonosis control programs.

In 1978, the Jaxx family, including two children born in Germany, moved back with the military to Kansas. Their six enjoyable years in the Army had changed their minds and they opted to stay in the service, which meant post-graduate training in biomedical research and development with the Veterinary Corps. In 1979 they loaded up the kids, dogs, cats, a macaw and a cockatoo and moved to Ft. Detrick, MD, to embark on challenging new career paths that would immerse them in Cold War biological and chemical defence programs. They would be engaged in the battle against emerging infectious and zoonotic diseases that would shape the rest of their personal and professional lives.

The US Army Medical Research Institute for Infectious Diseases (USAMRIID) at Ft. Detrick is one of the premier bio-containment laboratories in the world. It is the primary research organization responsible for developing medical countermeasures to protect US military personnel against potential bio-warfare agents. Most bio-warfare organisms are zoonotic (transmitted between animals and humans) and can cause serious disease. Thus specialized facilities, equipment and training are necessary for bio-defence research programs. Specialty-trained veterinary scientists such as pathologists, microbiologists, immunologists, virologists and laboratory animal medicine practitioners are required for the critical animal modeling and testing.

Despite ratification of the Biological Weapons Convention in the 1970s that banned the use of biological agents or toxins as weapons, the USA and its allies strongly suspected the existence of state-sponsored offensive biowarfare programs. Accordingly, the Institute's biodefense programs included a veritable who's who of nasty diseases, including anthrax, plague, botulism, tularemia and brucella. As well, there were exotic hemorrhagic fever viruses endemic to South America, Africa and Asia, including Junin (Argentine) Machupo (Bolivian), Lassa, Marburg, Ebola, Brazilian and Crimean-Congo hemorrhagic fever and Rift Valley Fever. Hemorrhagic fevers are often lumped together even though they are caused by different viruses because they cause similar effects in patients, including high fever with severe disturbances in the body's physiologic and blood clotting system that lead to shock, internal and external bleeding, and often death. With the exception of Ebola and Marburg, the location of the host animals is known, but the actual pathogenesis (method of disease development) of each agent differs, and most are still not well under-stood.

The Marburg virus was the first filovirus to be identified after an outbreak in Marburg, Germany in 1967 in laboratory personnel working with imported African green monkeys used for vaccine and serum production. There were 37 cases of Marburg hemorrhagic fever, with a mortality rate of 23 per cent. The Marburg outbreak is histori-cally significant because it was the first confirmed emergence of a filovirus (Ebola emerged in 1976 in sub-Saharan Africa). The outbreak occurred in a facility where excellent records of both human and non-human primates were maintained and the source was identified as African green monkeys from Uganda. At that time, African green monkeys were routinely imported for production of poliovirus vaccine in tissue culture. The emergence of Marburg heightened awareness of the potential for other unknown emerging diseases that might be present in non-human primates. In addition, because the outbreak was so lethal, it showed that the monkeys were not the reservoir for the filovirus, but served as an amplification host. Bats are suspected to be the most likely reservoir, but there is no firm scientific proof.

The Ebola virus is the only other member of the filovirus family and is the deadliest known hemorrhagic fever. There are multiple strains, but the Zaire strain has had the most profound public health consequences. Ebola Zaire has the highest mortality (more than 90 per cent in some outbreaks), and the largest number of human cases with more than 20 distinct outbreaks reported. Having only been seen in humans in explosive disease outbreaks that killed hundreds in Zaire in 1976, and the Sudan in 1979, Ebola was not thought to be zoonotic. The disease was largely unknown except to tropical medicine specialists and high hazard infectious disease researchers. The reservoir for the virus was completely unknown and there was no effective treatment.

When Jerry and Nancy arrived at USAMRIID in 1979, animal models were being developed and the search for potential antidotes and vaccines for the "the mystery virus" was in full swing. Nancy's supervisor, Dr. Anthony Johnson, asked if she was interested in working with Ebola and Marburg because the work suited her background in clinical medicine and interest in pathology. Nancy had some initial reservations because she and Jerry had two small children and the virus was exceedingly dangerous, but she accepted. Dr. Johnson taught her to function in the hot biohazard suits, carefully training her in procedures and the conduct of necropsies under high hazard containment conditions.

Most of the hemorrhagic fevers for which the host reservoir had been identified were either arena or bunya viruses, so it was natural to look at their reservoir hosts first. Exhaustive examinations of the flora and fauna where the 1976 and 1979 Ebola outbreaks occurred failed to identify the reservoir in the usual suspects, rodents and insects. None of these hosts fit the Ebola (or Marburg) picture, although bats were suspected even then. Worse yet, initial cases of Ebola had not been diagnosed, so the outbreak was in its third or fourth round before the culprit was identified. How the virus got into the human population was a mystery. Additionally, little about how the disease developed could be gleaned from autopsies with the diagnostic tools available at the time and it would be years before molecular biology caught up to the task. After the 1979 outbreak of the Sudan strain of Ebola, the virus "disappeared" for a decade, re-emerging unexpectedly in 1989 in Reston, Virginia in a way that would shock the infectious disease and public health community.

Ironically, after working elsewhere Jerry and Nancy had returned to USAMRIID in the summer of 1989 to become the chiefs of the pathology and veterinary medicine divisions. That fall, a veterinary pathologist at a commercial quarantine facility sent tissue samples from crab eating macaques. The monkeys had been imported from the Philippines and had died from a fatal syndrome consistent with simian hemorrhagic fever (SHF), a non-zoonotic viral disease of macaques. As expected, tests confirmed SHF. Results of additional tests sent a shock wave through our laboratories when they revealed that apparently the monkeys were also infected with a distinctive thread-like filovirus. Nancy confirmed that the tissue lesions were consistent with filovirus infection and she suspected the deadly Marburg virus because it was the only filovirus associated with monkeys in natural outbreaks. However, additional tests identified the virus as the one associated with the Zaire strain of Ebola. With the confirmation that a deadly hemorrhagic viral disease was spreading through a monkey quarantine facility in Virginia, a community so close to Washington D.C., the world's infectious disease research community was astounded.

Our leadership immediately consulted infectious and tropical disease experts, including those from the army, the Centers for Disease Control (CDC), World Health Organization (WHO), and the public health departments of Maryland and Virginia. Fortunately, several prominent filovirus experts were at the CDC at the time and they came to the conclusion that this was a dangerous threat to public health and that an extraordinary and unprecedented response effort was needed.

In the chaotic days that followed, there were more questions than answers. Ebola was an African virus, but these monkeys were from Asia; Ebola had never been reported in wild monkeys, but it was killing macaques in Virginia. How many persons had been infected as the monkeys were imported and how quickly would it spread to the surrounding community? Finally, what should be done to respond and contain the outbreak, and who would do it? The unsettling reality was that in late 1989 there was no established national plan or infrastructure in the USA to respond to a dangerous emerging disease such as Ebola. It was decided the CDC would take the lead for the complex public health response to the outbreak, while USAMRIID's Veterinary Medicine Division (VMD) was responsible for dealing with the contaminated quarantine facility and the Ebola-infected animals. Although no one organization had the specialized training or exotic equipment needed for this task, USAMRIID was best positioned with a contingent of experienced high-hazard animal care personnel and access to expertise and supplies needed to handle the quarantine facility.

About 48 hours after the confirmation of Ebola, VMD teams and safety and technical teams headed to the quarantine facility at Reston. The commercial facility was adequate for quarantine of monkeys, but was not capable of achieving the bio-containment conditions necessary for handling the Ebola virus safely. The biggest problem was the lack of a sophisticated filtered ventilation system and air pressure tools to help prevent airborne spread of the virus among nearly 500 monkeys in the facility. The monkeys were continuing to become infected and die in a pattern that suggested the virus was transmitted through the air. Although no humans had yet become ill, it was assumed cases were inevitable because routine handling and housing practices had been in use in the facility.

The first order of business was to set up an emergency bio-containment area. After carefully assessing the building configuration, entry and exit rules were established by designating cold, transitional and hot zones. Personnel from USAMRIID and VMD faced many challenges during the Reston response, but their overriding focus was on containing the virus, managing the safety of personnel, and delivering efficient and humane treatment for the monkeys.

Many memorable incidents occurred during the response, including the failure of the building's heating, ventilation and air conditioning unit, which sent the ambient temperature soaring to the mid 90s°F (32°C). On our first full day at Reston, one of the employees returned to the facility to gather some personal effects and unceremoniously fell to his knees in the grass and vomited repeatedly. This evoked a predictable reaction in our team and the public health personnel, but it turned out that he had a severe case of seasonal influenza, not Ebola.

The most problematic issue was the fact only about 40 per cent of the animals were housed in squeeze cages that greatly facilitate safe handling of monkeys. A safe method had to be developed to catch and anesthetize wild monkeys, including males with long sharp canines, in a way that minimized the chance of bites, exposure to the virus and escape from the cage. We eventually settled on using a modified pole syringe.

As Murphy's Law would dictate, we did have a monkey escape from its cage into an animal room. Retrospectively, this escape vividly contrasted the difference between the "Hollywood" army and the real army. In the movie "Outbreak", inspired by the Reston Ebola incident, the Hollywood army scientists decide a monkey must be carrying the exotic hemorrhagic fever virus that was infecting and killing thousands of people on the West Coast. They made a plea on TV, asking if anyone had seen the monkey. A woman had seen a crayon drawing of a monkey in her daughter's room and concluded her daughter must have seen it. The scientists flew to her home, threw some apples out on her deck and waited in the bushes. When the monkey appears out of a redwood forest, they shoot it with a capture gun and take it back to the command centre where they collect enough immune sera to fill a couple of tanker trucks, and save the world. What really happened at Reston was that the army had a monkey escape in the morning and three people chased it in a 12-by-35-foot room all day long before catching it. Although humorous in retrospect, it is significant that most people have the unrealistic expectations of an outcome as was depicted in "Outbreak".

After a hectic week, the Reston quarantine facility was depopulated and decontaminated. During that week, tissues from infected animals were continuously transported to USAMRIID's pathology and disease assessment division for further evaluation. It was eventually discovered that a completely new strain called Ebola Reston had emerged. Although apparently not fatal in humans, four out of five employees in the facility did develop antibodies to the virus. Had it been the Zaire strain with 90% mortality, several of these people and some of their friends would likely have died. As for our 42 team members, not one developed antibodies. In the view of the infectious disease community, this data verified that our emergency procedures and equipment worked.

As research capabilities have improved, it is possible to differentiate among the various strains of both Ebola and Marburg. To date, after 20 separate outbreaks, five species of Ebola have been identified. In 1994 the Ivory Coast strain was identified, a strain responsible for the decimation of chimpanzees that had been under long-term behavioural observation. Little more than a year later, the deadly 1995 Kikwit Ebola Zaire outbreak coincided with the release of Richard Preston's bestseller, *The Hot Zone*. What had been an obscure and rare virus became a household word virtually overnight.

Nancy Jaax's group at USAMRIID established that Ebola Zaire was easily transmitted via oral contact and through contact with the moist membrane in the eye. This provided scientific proof for how the virus could "jump" into the human population – non-human primates (bush meat) are commonly eaten in sub-Saharan Africa. USAMRIID scientists also established how the virus progresses and showed that infection with Ebola causes destruction of lymphocytes responsible for the immune response. Ebola profoundly and acutely suppresses the immune system, which allowed the virus to grow and replicate at an astounding rate. Further investigations led to immense progress in the development of vaccine and medical treatments for the disease, although there is no "silver bullet" treatment yet.

Observations of the Ivory Coast chimpanzee outbreak in 1994 supported the "food chain" theory of infection of people who eat non-human primates, as did subsequent outbreaks of the Gabon strain that decimated the gorilla populations in Africa from 1996 through 2003. Despite intensive and thorough investigation and sample collection in all natural outbreaks, the reservoir and natural life-cycle of the virus has still to be identified with certainty.

Richard Preston made an enormously powerful contribution to the cause of public health and infectious disease research when he published the book, *The Hot Zone*. For the first time there was a contemporary non-fiction book about a fascinating "mystery disease" that could be easily read by bright elementary school students, yet was still absorbing to adults of all educational levels. It put a face and personality on infectious disease researchers and led numerous young and bright minds into the fields of medicine, veterinary medicine, microbiology and immunology at a time when they were badly needed. Veterinarians owe Richard Preston a special debt; he showcased the significant roles and contributions that veterinarians make to human health in a way that was unprecedented. Preston went on to write other successful books focused on infectious disease and biowarfare topics, but none captured the public's fancy like *The Hot Zone*. For his contributions in raising the awareness of emerging and zoonotic infectious disease in a public health venue, Richard Preston is currently the only non-physician to be awarded a lifetime achievement award from the American Medical Association.

Postscript

At the heart of this story is an overview of the Ebola fever outbreak in monkeys in a quarantine facility in Reston. It is told from the personal perspective of a military couple, both veterinarians, caught up in the emergency response. This incident and the notoriety received from Richard Preston's New Yorker Magazine article, "Crisis in the Hot Zone", and subsequent New York Times No.1 bestselling book, *The Hot Zone*, would change our lives. Prior to his writings, few had ever heard of filoviruses or hemorrhagic fevers. However, *The Hot Zone* made Ebola a poster child for emerging disease fears. Preston's "virus thrillers" were a clear signal to a wide audience about the mounting public health dangers we face from emerging zoonotic diseases and the growing spectre of biowarfare and bioterrorism.

In *The Hot Zone*, Preston also described specialized career niches for veterinarians that are not generally well known. Because of Preston's writings and the wildly fictionalized movie "Outbreak" inspired by the Reston event, we are often asked by students and young veterinarians how we got into our careers. Accordingly, we have taken the liberty in this article to briefly describe our personal path.

The successful response to the Reston Ebola emergency was ultimately a testament to the leadership, dedication and teamwork of scores of military, government and civilian scientists at USAMRIID, the CDC, and elsewhere. In recognition of that, we apologize for focusing on our efforts and omitting the names of dozens who played critical roles in the success of the operation. Finally, we would like to recognize our military enlisted and non-commissioned veterinary technicians. They serve willingly and well in diverse and sometimes hazardous conditions. They are the backbone of our military veterinary services who selflessly work for all of us. We are proud of them.

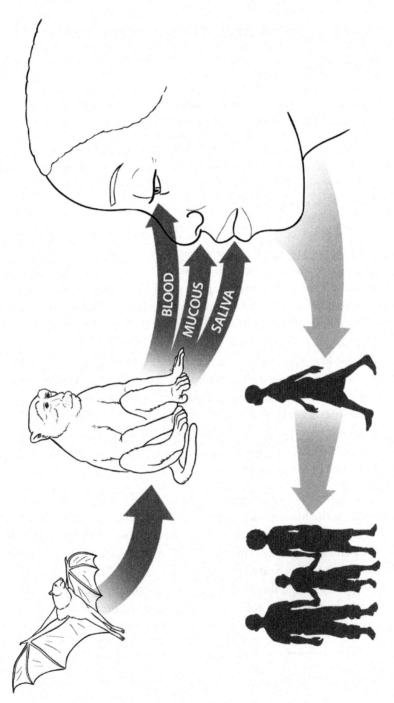

Figure 6 Ebola Virus

Dog Parasites in the Land of the Gods

David Waltner-Toews

Like many Westerners, I found enlightenment in Kathmandu. My gurus were a tapeworm and a Nepalese veterinarian.

Crossing the footbridge over the Bishnumati River in the chilly, dark predawn day in November, 1992, I saw a series of fires on the far bank, men and dark shapes of water buffaloes appearing and disappearing as the light flickered.

As our little coterie of scientists[1] drew closer, several men were grappling with one of the beasts, hammering a long, blood-covered spike into the base of the buffalo's skull. The buffalo sank into the straw as a man slit its throat to get a good blood flow. They piled straw over the body and fanned it into flames to burn off the coarse black hair. When the fire died down, they rolled the stiffening body onto its back, and slit open the belly. One barefoot man, his loose slacks rolled up to his knees, pulled out the huge and bulging white, blood-flecked rumen (stomach of a buffalo) from the abdomen. Another cut open the chest and took out the heart and lungs. Wading around in the spilled intestines, stomach contents and feces, steam rising into the chilled air, they quietly went about the work of preparing food for tourists and other wealthy consumers, while dogs and pigs nosed in and out of the growing piles.

Scanning the scene before me, I wondered about the big black lumps on the branches of the huge trees stretching up from the muck. Fruiting bodies of some sort, I assumed until one of them stretched its wings and glided away. These were vultures I realized with a shudder, hundreds of them, and a giant murder of crows, thousands of them. Our guide through this world of blood and shadows was Dr. Joshi, soon to be friend and colleague, a Nepalese veterinarian with a passionate dedication to improving the lives of his compatriots. He

---

[1] Dr. D. D. Joshi, Nepalese veterinarian; Dr. Dominique Baronet, a Canadian veterinarian and my graduate student; Dr. Peter Schantz, veterinary parasitologist from the United States Center for Disease Control; Dr. Don de Savigny, epidemiologist and parasitologist from the Canadian International Development Research Centre (our funder); and the author

loosened the scarf around his neck, tucked back the sleeve of his white shirt, and pointed to the riverbank beside us, where piles of buffalo feet, manure, blood and offal spilled down into the small black trickle, which was all that was left of the river. Ducks, pigs and dogs foraged for maggots and bits of discarded bone and fat, and, down by the trickle of water, men and women defecated and washed.

"There used to be broad, stone stairs down to the river here," Joshi said. "Somewhere underneath all that . . ." he waved his hand. "People would go down for their morning ablutions."

We stepped away from the main slaughtering area, up the steeply sloped alleys, past several groups of buffaloes awaiting the knife. Kids clambered over them; this was their playground. We watched dogs flow toward the slaughtering areas and collected dog feces in little film containers, marking the locations on our tourist city maps.

A rickshaw being pushed uphill past us had a pile of animal parts spilling over its sides. Dr. Schantz pointed out a fist-sized, whitish, semi-opaque cyst in a buffalo lung. This parasite cyst, called a hydatid cyst, was the reason Dr. Joshi had brought us to this medieval city that was his home. These fleshy balls can be found in many kinds of herbivorous animals, which serve as the intermediate host for the parasite, with sheep being the most important worldwide.

Hydatid cysts also occur in people, where they can cause severe disease. Cysts can occur almost anywhere in the body, but usually end up in the liver, where it acts as a slowly growing tumour and causes abdominal discomfort. In severe cases, people look as if they are pregnant; and in an alien-species-type way, they *are* pregnant with thousands of tiny tapeworms. If the parasite makes a cyst in the lungs, it can cause a dry cough or, if it ruptures, coughing of blood. Surgery to remove the cysts can be risky because rupture of the cyst will send hundreds of tiny "daughters" throughout the body, causing shock and death. In a survey of local hospitals, Dr. Joshi found that more than 20 per cent of people who had surgery for cyst removal died during or shortly after surgery. Anti-parasitic drugs, which have been tried in various combinations and for different periods of time, are only partly effective in reducing cyst size and require long-term treatment.

People, like other intermediate hosts, get infected by inadvertently eating the eggs, which are found in dog poop. But when the unfortunate tapeworm ends up in a human, it finds itself stuck as a stranger in a strange land. Because people are not (usually) eaten by dogs, infected people are what parasitologists call a dead-end host.

In dogs, the parasite *Echinococcus granulosus*, is a tiny, mostly harmless tapeworm. If you carefully, painstakingly finger your way through piles of fresh dog feces (much to the delightful scorn of neighbourhood children), or better yet, from an investigative point of view, thread your way through the intestines of a freshly dead dog, you can sometimes find what looks like tiny, glistening bits of raw pasta. These are the adult tapeworms containing embryonated eggs, which are released into the environment.

Dogs (and coyotes and foxes) get infected when they eat the parasitic cyst, such as the kind we had seen in the buffalo lung in the rickshaw; people could eat the cysts without becoming infected, though I don't know of any good recipes offhand. Once the dog ingests the cysts, hundreds of immature tapeworms (protoscolices) within the cyst mature into tapeworms that live in the intestine. The tapeworms are good guests in the dog, causing little disease or discomfort. From this safe, warm, home, the tapeworms release eggs into the dog poop, starting the life cycle over again.

That day in 1992, having seen the animals being slaughtered, the cysts and the dogs, we continued up the narrow brick and pot-holed asphalt streets between ancient brick walls and finely carved cracked, ancient wooden balustrades. We circled back down the main road to the Bhimsentan Bridge. The vendors were laying out their wares on the sidewalk: brass pots, cotton shirts, and fruit. Tailors were tuning their sewing machines. Below, the pigs foraged in offal from slaughtered buffaloes and open latrines. Nearby, a girl on the sidewalk carefully stacked lychees and peaches in pyramids of five. With nimble grace and dignity she crossed her legs and waited.

By the time we walked back to Dr. Joshi's place, the sun was up and the temperature rising quickly. By mid-day it was in the mid-20s C and we were down to T-Shirts. We gathered around a wooden table in a small room at the National Zoonoses and Food Hygiene Research Centre. The Centre was Joshi's modest two-storey house with a large black-and-white sign at the gate. I could almost imagine that we were a revolutionary cell of scientists ready to bring hope, enlightenment, clean water, less garbage and an animal inspection act to the dark chaos around us.

We knew that to understand this parasitic disease we would need to do surveys of infection in livestock, dogs and people. We would need to understand something of livestock movements because many of the animals being slaughtered in the valley came from elsewhere and hence would have been infected elsewhere. We would need to understand something about how people and their dogs relate to each other. In some ways, it was a relief to have all these details to focus on. And if we understood all these things could we not also make them better? Didn't knowing the problem take us halfway to having solved it?

That evening, I stopped my bicycle in the middle of the bridge to gaze up the valley. I recalled a visit earlier that week with one of the keen young officials in the Ministry of Health. I was there to inform him that we had some human rabies vaccine available, not just for our researchers, but for use by Ministry of Health personnel in the area. He proclaimed that Nepal had its own perfectly good vaccine and that the human diploid vaccine recommended by the World Health Organization caused cancer. He seemed annoyed that we (outside scientists) were starting a research project when the answers were already obvious. The way to get rid of both rabies and hydatid disease, he said, was through mass gassing of street dogs. I did not have the temerity to ask how this would be accomplished. I was, after all, a visitor, and his implication was that scientific imperialism had somehow insinuated itself into the niche left open by the political imperialists. The accusation had a ring of truth, and I was left feeling angry and disheartened. How did one demonstrate global solidarity without falling into the old patterns?

Still, we were determined to save this area (wards 19 and 20) of Kathmandu. And if we could reclaim this small part of Shangri-La from the depths into which it had fallen, was there not hope for the whole world? Might there not be a path into a sustainable, healthy, convivial future?

***

In June of 1995, as I flew into Kathmandu from Bangkok, my mental landscape was in turmoil. Why hadn't the work of our research team made any discernible impact?

The science part of this equation involved gathering evidence. The applied part was to convince policy makers to take up the knowledge we generated and use it. We had certainly gathered evidence.

Since we had begun our work in 1992, Drs. Baronet and Joshi and the other researchers doggedly pursued the demands of good, conventional science. Working with our Nepalese colleagues, Dr. Baronet documented the behaviour of dogs in Kathmandu and collected fresh dog feces. She then put fluorescent collars on the dogs, took pictures of them, and gave vaccines for rabies and anti-parasitic drugs. She developed a veritable mugs gallery of the street dogs of Kathmandu. She not only probed through dog feces searching for those rice-grain-sized tapeworm segments but also, working with a new laboratory test, searched the feces for bits of tapeworm too small to see with even the best microscopes.

The picture that emerged from our evidence was complicated. There were cysts in five to eight per cent of the buffaloes, goats and sheep slaughtered along the riverbank. Many of these animals came from India and Tibet, so the primary cycles of infection were probably in those countries. Of the dog owners in the areas where livestock were slaughtered, about 60% fed their dogs raw meat, offal, bones or cysts as their main source of protein, but only about six per cent of the dogs in Kathmandu were infected and re-infection rates were very low, suggesting that the disease was primarily imported.

We discovered that infections in both dogs and people also occurred outside of the butchering areas. The butchers did not cut the hydatid cysts out of the meat, but instead sold them along with the meat that was priced according to weight. Once the squeamish consumers got home, they cut out the cysts and fed them to their dogs. Working with the veterinary clinics we had recruited throughout the city, we were able to identify dogs in "suburbia" that were infected in this way. Many of the dogs were valued as pets or guard dogs. People got infected by getting trace amounts of dog poop on their hands, and then eating or smoking.

We had hoped that the epidemiological studies would, through the sheer weight of evidence, convince people to change their ways. The solutions were, on the face of it, obvious: 1) have health inspectors at the slaughtering areas cut out cysts and dispose of them properly; 2) close in the slaughtering areas and educate the public so that the dogs didn't get access to raw, cyst-infested meat; 3) educate people about hand-washing after handling their dogs and about cleaning up dog feces to decrease fecal contamination of streets and houses; 4) control the dog populations. Dr. Baronet prepared an educational video in Nepalese that was shown on television. Dr. Joshi ran public education clinics and talked to members of the government about legislation. Yet, after several years of hard work and good science, nothing seemed to have changed. What was missing was any deep sense of culture: why did people behave in certain ways and not in others?

Scientific information is assumed to be objective and globally true, but can there be a science of place, one that explores phenomena that are locally true?

Even such a simple thing as having a national act to regulate animal slaughtering, which Dr. Joshi had been working to get enacted for years, was stalled by political events. Nepal seemed to have moved rapidly from a system of the King's rules to a system of no rules. Now, when I bicycled around the crowded valley, I saw that the once-green rice fields were strewn with new, hastily erected and already crumbling buildings. Everything was now for sale. All this economic activity was bringing in a lot of money. Many politicians, as well as the slaughter-house owners and butchers, were responding to the demand for meat from tourists and the newly wealthy in this free-for-all economy. Could they be convinced to reinvest in the health of their communities? How could we engage them in the work so that they would see its importance? My view of the relationship between science, policy and public practice was shifting.

I thought about the web of connections, as well as the facts, that our research had uncovered. They explained a lot about why some previous attempts to improve the well being of the urban citizens in Kathmandu had been unsuccessful. Family and cultural traditions of butchering, perhaps once sustainable in sparsely populated rural areas, were brought into the crowded urban setting of Kathmandu. Butchers didn't want to give up family traditions and become wage labourers in an animal-killing factory.

Whatever would be done to prevent infections with this parasite would have to be embedded in a much more comprehensive program of social transformation. If ecosystem approaches to health were a viable ways of thinking about and working with complex public health problems, Kathmandu seemed to be an ideal case study.

At dusk, I sat nursing a beer on the roof terrace at the wonderful old Vajra Hotel, with its wood-carved walls, gardens and traditional statuary, on the hillside up toward the monkey-temple. I looked out over the darkening valley, the scattering of lights in this still-medieval city, the satellite dishes along the rooftops. The scent of dung and wood-fires and the clashing sounds of popular Indian and European music drifted up past me and into the cool almost clean mountain air. If these new "ecohealth" ideas about the health of social and ecological systems, combining community participation with systemic ecological understanding, could work here, they could work anywhere.

Postscript

People often ask me what happened to hydatid disease in Kathmandu. I don't know, in part because it wasn't a priority for the citizens of wards 19 and 20. I do know, however, that the conditions that had made it a problem have changed for the better.

In November 2001 we had the final workshop of a community-based ecosystem health project. Our research team this time had included butchers, street sweepers, veterinarians, politicians, shop owners, anthropologists and community activists.

Later, I took a picture of Dinesh Kadji, dynamic young chairman of the butchers' association, standing outside in the clean, sun-lit courtyard beside the river. I told him that I thought there used to be an old water-pump here. I had a picture of a woman taking water, with her children squatting around her, and, in the background, animals being slaughtered. It's right over there, he said, pointing to a clean, well-kept building across the yard, and laughed.

The appalling riverbanks I had first encountered in 1992 were transformed. Public parks with shrubs and trees and benches to sit on had been created next to the Bhimsentan Bridge. A man and a woman took care of new public toilets, as well as the gardens. Although people were prohibited from defecating at the riverbank, I could see a fellow coming up out of the riverside grass, pulling up his pants. Still, it was much cleaner than it had been a decade previously. There were some piles of garbage, but many fewer than before. There were fewer dogs as well. The trees were empty of vultures.

Next to the public parks, further along the riverbank, were walls of tall grasses, and small, private flower and vegetable gardens. Beyond that, rows of composting cow manure and stomach contents were aerated by plastic pipes; ducks, picking at bugs and grubs, wandered between the rows. To my right, away from the river, were cardboard and plastic recycling yards. A row of water buffaloes was walking slowly along the street, followed by one lone herder. The herder guided them up a side street to a corrugated metal door. I slipped through the door after the animals. Around a brick courtyard were covered sheds. This was one of the 20 or so new, mid-scale slaughter-ing places scattered throughout ward 19 and 20. At a fresh-water tap, women and children were pumping water into their brass pots and plastic pails.

Back outside, along the side streets, I came across the perpetually running stone taps. These taps are set in washing places below ground level for general community use, and are maintained by traditional religious caretakers. Once filled with garbage, some of these washing places were now cleaned out and painted and were being used.

At one of our research workshops, the participants were asked what stories they wanted their children to tell about them. They said that their actions in response to the research were both the stories and the evidence that would be their legacy.

What I hear from Dr. Joshi is that the communities are continuing their work, fighting for their democratic rights, creating stories of resilience in the midst of the battles among the capitalist, royalist and Maoist ideologues. They have been monitoring water quality. Joshi himself is still running dog vaccination clinics, and recently received a prestigious award from the Nepalese government for his contributions to national science and well being. The social-action Nepalese partners

I worked with have mounted programs to improve nutrition and health of the people in ward 19 and 20.

The people there have discovered the story they want their grand-children to tell, and it is one of local democracy and commitment, of no tapeworms and of clean water and a thirst for knowledge, of green riverbanks and impatience with systemic poverty and repression. I often think back to 1992, and remember the small girl on the bridge, waiting patiently with her lychees and peaches, and I hope she has a happy role in that story.

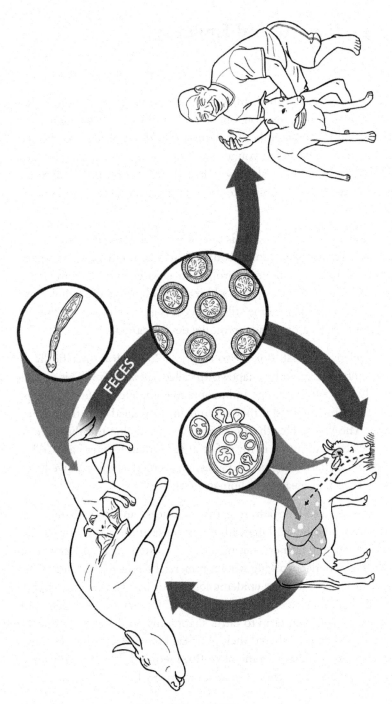

Figure 7 Hydatid Disease

Pigs, Poverty and Epilepsy

Cate Dewey

His wide, white-toothed smile and trusting eyes look up at me. He is waiting for some attention, one child in a thousand, longing for a hug, expecting I will remember his name and take his photo. His face is covered in deep, angry wounds, surrounded by swollen, tender tissue, scabbed over and just starting to heal. It reflects stark evidence of a grand mal seizure, falling face down in the dirt and mud and gravel and sticks.

Rodger has epilepsy. He is one of many children and adults afflicted with this dreaded clinical condition in a community where epilepsy is poorly understood, never properly diagnosed and rarely treated. In this culture, old superstitions add fear and blame to the families of those with epilepsy. Some still believe it is a punishment for the sins of a mother, father or grandparent, the devil incarnate.

Epilepsy is what took me to Kenya in 2006. More specifically, it was epilepsy caused by a tapeworm called *Taenia solium*. This parasite needs to live in both pigs and people to complete its life cycle. Pigs only become infected if they eat human stool filled with tapeworm eggs. The World Health Organization suggests that this disease can be eradicated and, sitting in my office at the Ontario Veterinary College in Guelph, Ontario, I agreed. Eradication seemed simple. All we have to do is keep pigs away from human stool.

I traveled to Kenya to research the disease and immediately realized that we would be working with the poorest of the poor. For these subsistence pig farmers, raising a pig was a step out of poverty. I had a lot of experience working with commercial pig farms in North America as a veterinary epidemiologist, but I wasn't sure my expertise would help these Kenyan farmers. Over 30 years as a veterinarian, I have traveled from farm to farm, sharing my book knowledge with the farmers while they shared their practical knowledge. What one farmer teaches me, I share with the next. But I worried that the Kenyans might not accept a white, Canadian woman who couldn't speak Swahili.

On my first day in the field, we were walking down a path from one farm to another and suddenly faced a cloud of blue moving towards us. Nearly 300 school children wearing blue uniforms were walking home for lunch. The children pointed at me and yelled "Mzungu, Mzungu." My Kenyan friend told me this Swahili word means white person. Most of the children stared and gave me a wide berth as they passed, keeping their eyes glued to my face. Three boys, who must have been especially curious, stopped directly in front of me. They looked about 10 years old. The one in the middle was clearly the brave one and he said "Mzungu, how are you?" and held his hand out for me to shake. Just as I was going to shake his hand, he pulled it back, as if my hand would burn his. I squatted down to be at his level, and then said "I'm fine, how are you?" I kept my hand extended. His courage returned and he shook my hand and then stepped back and puffed out his chest, looking very proud of himself. Next, his two friends shook my hand. My Kenyan colleague told me that they would tell their grandchildren that they shook the hand of a Mzungu. You see, these children had never seen a white person before.

The farmers were just as accepting. They were living in abject poverty, and yet were filled with love and joy and welcomed me, a total stranger, into their midst. It was the first time anyone had come to help them with pig keeping. The pig was a way for families to bank their wealth. If there was a little money from the maize (corn) harvest, they bought a piglet. The pig was raised by the woman in the home. She sold the pig when there was a family emergency, to pay for medicine or a hospital stay, school fees or food between harvests.

Subsistence farmers hoe the land by hand and pray for a fruitful combination of rain and sunshine. If the rains come, the harvest is good and there are maize and beans to eat. When the rain doesn't come, the harvest is small. I arrived in Kenya in February 2006; this was the third successive year of drought and this one was severe. The harvest was poor. The farmers, living on less than one dollar a day, could not afford to buy food. People were starving. Children sat listless in the sun next to their mud hut, their bellies swollen, their eyes watching as I worked with the pigs.

Pigs eat the same food as people and compete for food with the children. The farmer's option is to feed the pig or feed the children. The solution for most farmers is to let the pig roam freely to scavenge what food they can find. Lucky pigs, living close to village centres, eat waste from the market and offal from the butcher shop and spend hours nosing around in garbage heaps. Most pigs eat fallen fruit, vegetable peels, weeds, the neighbours' crops and human stool. Walking from farm to farm, meeting these people face to face, made me see their reality. The eradication of *T. solium*, the most common cause of epilepsy in the developing world, seemed far from possible.

The farmers live in mud huts they have made themselves, with only a crude outhouse with a hole in the ground to collect the urine and stool. Farm families have no running water and cook their meals over a wood fire. Daily chores for women and children are to gather wood and water. Homes are filled with children, often with a single woman having eight to 14 in her care. I asked about these large families and was told half of the children in the villages are orphans because their parents have died of HIV-AIDS. Grandmothers who had buried their own children are raising their grandchildren. Women and men who had buried all of their siblings are raising their nieces and nephews. Some children have no living relatives, so they are taken in by neighbours. I wondered if I would do this. If I had no money and not enough food for my own three sons, would I take in eight orphaned children knowing that my children would go hungry? I was struck by a new reality. This is a generous culture. Everyone accepts the fact that the children need to be cared for and are welcomed into a home in the village.

Although many people in this area of western Kenya develop epilepsy from tapeworm disease, no one was aware about the connection between the tapeworm, pigs and epilepsy. The government livestock and veterinary extension officers who mentored the farmers did not know, and neither did the pharmacists, nurses, public health advisers, farmers or pig butchers.

I showed them the connection - people get this tapeworm by eating undercooked pork infected with tapeworm cysts. The tapeworm then matures into an adult in the person's bowel. A few weeks later, the adult tapeworm begins laying 50,000 eggs a day. The tiny microscopic eggs pass through the bowel and are shed in the person's stool. When a pig eats the egg-laden stool, tapeworm larvae hatch and migrate to the pig's muscle where they develop into a larval cyst. Anyone who eats undercooked pork containing larval cysts can become infected with a tapeworm that will then shed eggs in the stool. Thus, the life cycle of the tapeworm continues.

The typical life cycle is disrupted when a person ingests a tapeworm egg which has been passed in their stool or another person's stool. In this situation, the larva doesn't stay in the bowel; instead it migrates to the person's brain and forms a larval cyst. The pressure of the cyst in the brain causes epilepsy.

Farms and homes in the village usually have no source of water. Women and children walk 2 to 15 kilometres each day carrying jugs of water on their heads. Water is precious, making personal hygiene difficult. If a person doesn't wash their hands after defecating, they could easily consume tapeworm eggs without knowing because the eggs are so small that they can't be seen.

Visiting the homes of people with epilepsy made me realize the social stigma attached to the clinical problem. Filled with shame, many families keep those with epilepsy at home. For example, Rose was a 28-year-old woman who had never left her farm and was terrified of our research team. She hadn't gone to school, will never work and never marry. It is rare for people with epilepsy to get treatment because the hospital is too far away and there is no money for medication. Families that do sell a sheep or a pig to pay for a doctor's visit and some medication tend to give up treatment as soon as the person has a seizure while on the medication. They don't understand that the pills will lessen the frequency of the seizures but not provide a cure. I was desperate to teach the community how to prevent tapeworm infections and the concurrent epilepsy problems. Even if I could prevent one person from developing epilepsy, it would be worth the effort.

I faced another reality as I traveled from farm to farm – it was impossible to tell the difference between the orphaned children and those who lived with their parents, except at noon when school children came home for lunch. Alice and her husband cared for 12 children in a mud hut and had six pigs and a collapsed outhouse. With these factors and a shortage of water, I worried that someone was bound to become infected and develop epilepsy. Our research team visited their house just as Alice's four sons arrived home for lunch from school. Their uniforms were torn and ragged but still they were getting an education. The eight other children hadn't gone to school; these were orphans including two granddaughters, three nephews and three neighbours.

When we finished handling the pigs, Alice brought us some of her precious water to wash our hands.

"Why aren't all of the children going to school?" I asked her.

"I can't afford pencils and uniforms for the orphaned children," she answered.

Obviously the farmer's money was needed to buy seeds to plant the fields for food to feed the 14 hungry mouths. But it wasn't just this family; this was common throughout the whole community. Where was the future of this community when half of the children were orphaned and not going to school? This question was the seed for a new outreach project aimed at helping these subsistence pig farmers educate children who were AIDS orphans.

Although I was certain that education was the first step, how could I, as a Canadian veterinarian and professor, make this happen? Surely a few uniforms and pencils couldn't cost that much money. Maybe between my friends and family we could send some of these AIDS children to school. The emails I sent to Canada were received with enthusiasm and even some pledges for support. The original goal was to raise $15,000 over six years to support the children. During this time, we would work with the village to find ways for the community to continue supporting the children once our financial aid ended.

Armed with the promise of donors, I made my first official visit to the Bukati Primary School in the village of Butula to meet the school staff and community leaders and offer our assistance. Our hosts were pleased that we were willing to help and agreed that many AIDS orphans were not being educated. However, they said the real need was for a lunch program. For much of the year, food was scarce. In this third consecutive year of drought, children were dying of hunger. The adults in the community could not feed that many mouths. Even if the children did attend school, they had little chance of learning if they hadn't eaten. I was told that if we were to provide a lunch program, the children would not only come to school but all the children would also be able to learn. I felt as if I was being asked to do the impossible. The lunch program would cost hundreds of thousands of dollars. That was too much responsibility for me to accept.

At the same time, over the next two weeks, I experienced wonderful highs and dreadful lows as we continued to conduct our research. I was learning more and more about these farmers and the struggles they encountered as pig keepers. They told me that they couldn't tell how much the pigs weighed. As a result when they negotiated with the butchers over the value of the pig, they were at a disadvantage. Most of their pigs were thin, dehydrated and covered in lice. Farmers who had sows struggled to get them bred, and when they did give birth, only four to six piglets were born. As a veterinarian, I could teach the farmers to enhance their pig-keeping business and to improve the welfare of their pigs. I felt grateful for my veterinary education.

While I was becoming enthused about my plans for research and farmer outreach, every day I was witnessing more poverty, more orphans and more need. I knew that if I walked away from these orphans, no one else was likely to come along in time to help them go to school. I was the first 'Mzungu' they had ever seen. Finally, after two agonizing weeks of soul searching, I agreed to do my best to raise the money for a lunch program.

In July 2006, my family and I visited the school to measure the orphaned children for uniforms and give them pencils. Of the 580 students in the school, only 150 were AIDS orphans – even more proof that the orphaned children were not going to school. The school children presented plays and poetry, singing and dancing with a theme of AIDS prevention. Even the Grade 2 children sang about the pain of watching their parents get sick and die and the loneliness and despair of seeing them laid in the ground. The orphaned children lined up to be measured for uniforms by the local tailor. The girls lined up in rows by height; extra short, short, medium and tall and then the boys. The tailor measured the first girl in the row and then counted the girls in the row; 22 of 'this' size, 18 of 'that' size, and so it went until he had accounted for all of the orphans. The principal showed us the list of children's names. Beside each name was the child's age and grade, the year their parents died of AIDS and who was currently caring for them. The list went something like this: Kevin is 12, his parents died in 2002, he is being cared for by his 79-year-old grandmother who has eight children to feed; Evans is 10, his parents died in 1998 and 2004, and he is being raised by an aunt who has 14 children in her care. The principal told me that to get on this list of orphans, the children bring their parents' death certificates to school. For the Kenyans, this seemed matter of fact, but for this Canadian it was emotionally overwhelming.

The major focus of our research work was education as the means for empowerment. I invited my new Kenyan friend, veterinarian Dr. Florence Mutua, to be my partner in this work which evolved into a multi-year, multi-centred project. The project became Florence's graduate research project for her PhD and she in turn became my cultural guide and the essential key to unlocking the doors to rural Kenyan villages.

Our research project was conducted over four years in 52 villages, visiting 287 farmers every five months for three visits. During each visit, we weighed and measured each pig and developed a size/weight chart that was specific for the local breeds of pigs, matching the length and girth of a pig with its weight. The farmers told us about the importance of the pigs in their lives, how they bought and sold, fed and bred, and cared for their pigs. We found out that some provided no food for their pigs because the pigs just ran free. Others tethered their pigs and fed them a variety of locally available food. Slowly, we built a list of what pigs were consuming - weeds and leaves from sweet potatoes, rotten bananas and mangos, waste from the abattoir and the corn grist mill, mash from the beer made by the women at their homes and ugali, the cooked ground maize that is the staple food in the area. We came to realize that not all pigs were competing for food with the children because at least a portion of each pig's diet was not suitable for human consumption. What we learned from one farmer, we shared with the next farmer.

From this knowledge, Florence and I taught the government live-stock, veterinary and public health extension officers, nurses, social workers and adult education specialists about epilepsy and the lifecycle of the tapeworm and also pig housing, feeding, breeding and parasite control. We also showed them how to measure the length and girth of the pigs and use the size/weight chart to estimate the pig's weight before the farmer takes their pigs to the butcher. Then we facilitated these leaders to teach more farmers using the local Luhya language and Swahili. The team moved from village to village, setting up chairs under mango trees, using flip charts attached to fences. If a farmer couldn't attend the workshop, Florence or I traveled to their farm to give a one-on-one lesson (of course I needed a translator). Each farmer received a tape measure, a chart to track the weight gain of their pig, and two pictures, one of the life cycle of the tapeworm and another of all the foods being fed to the local pigs.

The ongoing research visits provided opportunities to work with the Bukati Primary School as it attracted new students and established the projects to ensure sustainability. The school community started with chickens and pigs and then added vegetable gardens, a tree nursery and a maize field on six acres of rented land. The school lunch program grew slowly. All the kindergarten children were fed five days a week from the beginning, but the rest of the orphans were only fed one day a week until we raised enough money to feed them on two, then three and finally four days.

In 2009, the Children of Bukati project purchased 11 acres of land for the school (www.childrenofbukati.com) and then developed a permaculture project on the land. Permaculture is a system of cultivation for agriculture, horticulture, and agro forestry that relies on renewable resources and a self-sustaining ecosystem. For the Bukati project, fish ponds, trees for nuts, fruit and lumber, rice, maize, beans, green vegetables, tomatoes and a tree nursery were all part of the system. With the sheep, dairy cattle, pigs and poultry, the goal of sustainability is becoming a reality. The permaculture project educates the community about enhanced agricultural practices using composting to improve soil quality, forestry and intercrop planting, and greenhouses and swales to lessen the effects of drought. Using food from the harvest and revenue from selling the excess, all of the children at the school are now eating lunch five days a week.

By 2010, there were more than 1,000 students at the Bukati Primary School. Of these, 750 were AIDS orphans, five times as many orphans than in 2006 and every orphan in the community of Butula was attending school. Even Rodger, who under ordinary circumstances would be kept at home because of the shame of his epilepsy, goes to school regularly. Instead of being an outsider, he is considered special. When he feels a seizure coming on, he tells his teacher Pamela who then takes him out of the classroom and stays with him during the episode.

One can see how the Bukati experience leaves me with deeply conflicting emotions. Our efforts have clearly demonstrated how a community can be strengthened and tapeworm-related epilepsy can be reduced and managed, a very positive outcome. At the same time, I recognize just how much needs to be done to help the poorest of the poor in Africa.

Our research work touched a relatively small number of farmers, butchers and government officials in one small corner of western Kenya. Although we are convinced that knowledge leads to action, it is difficult to know how to spread the message about preventing epilepsy caused by *T. solium* more broadly. Smallholder pig farmers across Africa, and indeed in many developing countries around the world, continue to let pigs roam freely, enabling the parasite to complete its life cycle, and to also get side-tracked into people's brains and cause more and more cases of epilepsy. They will join Rose and Rodger and so many others and live with the social stigma and physical debilitation of untreated, progressive epilepsy.

Even more heart wrenching are the AIDS orphans sitting in the sun outside a mud hut, their bellies swollen with hunger, staring out of hollow eyes without a whisper of hope in their hearts. My research takes me to so many communities that are suffering just as the Bukati school community was in 2006. Village elders plead with me to bring the Children of Bukati project to their local school, feeling it is so unfair that Canadians are helping Bukati while ignoring their children. There is so much hunger, so much need and an atmosphere of hopelessness.

Postscript

In 2010 and then in 2011, the Children of Bukati charity began assisting two more schools, the Bwaliro and Buduma Primary Schools, by first developing permaculture projects at both schools. Cate continues to visit the schools to meet with the children, staff and community and to ensure continued progress towards self-sustainability. It has been wonderful to watch the children grow and blossom year by year. However, in 2013, when Cate asked about Rodger, she was told that he had died of epilepsy. He would have been in grade 6. The Bwaliro Primary School is progressing well and was able to feed the children using produce from the land during one semester after only 3 years of development. The numbers of AIDS orphans and destitute children attending that school have increased from 170 to 460. These projects give the communities new hope, energy, and optimism for the future.

Today, the farmers use different pig management to keep themselves and their community healthier. Most pigs are either kept tethered to a bush or confined in small, homemade barns. This means that the farmers are carrying water for the pigs each day and collecting weeds and sweet potato vines and fruit to feed the pigs. Many farmers are providing pigs with a more balanced diet by buying small fish or butcher waste for protein and waste from the corn grist mill or ugali (corn porridge) for energy. The farmers use the measuring tape to estimate the pig's weight before it is sold to ensure the best price. With more feed, the pigs are bigger and healthier when they are sold. The villagers are selectively buying pork from the butchers rather than from a neighbour's backyard to ensure the meat is inspected. These significant behaviour changes will eventually reduce the prevalence of the tapeworm and the devastation of epilepsy in these villages. The lifecycle of the tapeworm will be broken because pigs are being kept from eating human stool and infected pork is not being eaten. Education is the key.

For more information on the Children of Bukati project, see www.childrenofbukati.com.

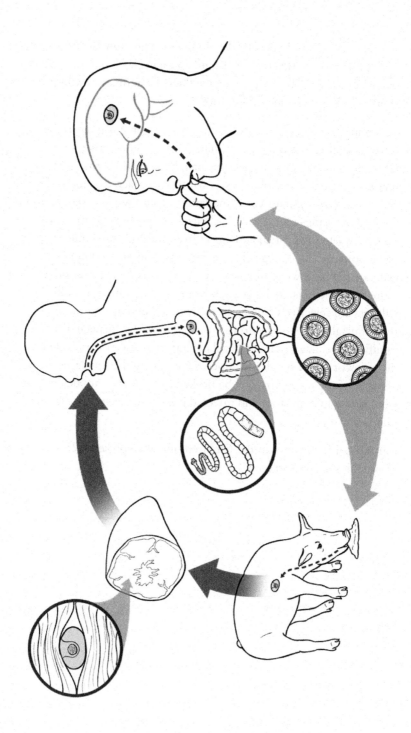

Figure 8 Tapeworm Disease

Sleeping Sickness Rebounds in Africa

Lea Berrang Ford

Imagine living in a small African village and your grandmother is ill with an unknown disease. Your family assumes it is malaria or AIDS and all you can get from the community health worker is the promise of malaria medication next time supplies arrive. You don't know what the problem is and she can't afford to go to the nearest hospital.

As you struggle with the grim reality that your grandmother is getting sicker and sicker, you see a bicycle coming along the rutted lane towards your small village. Curious, you step outside to see this stranger and discover he is a sleeping sickness control officer. Slightly encouraged, you tell him about your grandmother's symptoms. He listens to what you say, takes her to a Sleeping Sickness Treatment Centre and returns with a diagnosis of sleeping sickness (SS). She can be treated, but she has progressed to the late stages of illness; now she needs to be treated with strong drugs and her chance of recovery is less than if she had been diagnosed earlier. You feel guilty for not finding the money to take her for medical care earlier. But even if you had gotten her to a hospital, there is a good chance she would not have been diagnosed on her first visit.

Similar scenarios occurred frequently in Uganda in the 1960s and early 1970s as aggressive sleeping sickness programs were being introduced. I learned about stories such as these when I was doing fieldwork in Uganda many years later in 2002 and 2004. During my visits, I had many conversations with Dr. Dawson Mbulamberi, who had been a young physician during that time when he championed the idea of sending out special sleeping sickness officers to help monitor and address the disease. In that program, officers were paid a wage and got a bonus for every patient they brought in. Bicycles were a cheap and effective way to get around the countryside and the solution worked. Indeed, when I heard Dr. Mbulamberi describe the bicycle program and how it worked, it painted a picture in my mind of the impact sleeping sickness had on the people of Uganda.

In 2002 and 2004, my fieldwork was focused on why sleeping sickness had re-emerged after being mostly controlled three to four decades earlier. When I asked expert observers about what they thought caused sleeping sickness to re-emerge in Uganda, they would often answer, "Idi Amin". President of Uganda from 1971 to 1978, Idi Amin became president through a coup d'état against the previous president. Now known as the "butcher of Uganda", his brutal administration killed an estimated 100,000 to 500,000 Ugandans during his reign of terror. In step with the violence, the country's economy crumbled, with inflation reaching as high as 1,000 per cent.

For my study, I was using advanced spatial analysis to analyze changes in vegetation and land cover. These changes did prove to be important in explaining which areas were at risk for sleeping sickness. The belief by some of my colleagues that Idi Amin's regime also contributed to the re-emergence of sleeping sickness was both interesting and unexpected for me. This political component, the "Amin factor", led me to also study the role of conflict as a determinant of the disease. Since many intervention measures are premised or partially reliant on political stability for success, it is critical to recognize that sleeping sickness occurs in poor countries during and after a period of instability.

There are two types of sleeping sickness - Rhodesiense, in the east and south of Africa, and Gambiense, in the west and central part of Africa. Uganda is the only country with foci of both forms of the disease, but in the past their distributions had not overlapped. Sadly, this is changing. As sleeping sickness has re-emerged and spread, the two forms have moved closer together in central Uganda. Further complicating the sleeping sickness situation is the fact that the diseases are difficult to differentiate without advanced diagnostics (not readily available in rural Uganda), and they are controlled and treated differently (and some sleeping sickness drugs are quite toxic).

I once attended a meeting with international experts and coordinators of African sleeping sickness regarding prioritization of sleeping sickness control in Uganda. There had been some excellent GIS mapping (Geographic Information System, a system that stores, analyzes, captures, manages and presents data that are linked to location). The analysis had identified "priority" geographic areas for intervention. Despite this rigorous science and methodology, the identified areas were not consistent with what I was hearing on the ground. This puzzling result was due to the important – but often neglected – role of factors that are difficult to quantify and model, such as 'future' risk, war or conflict and community mobilization. Conflict of varying intensity has been ongoing in Uganda, even after the end of the civil war. Conflict leads to movement of people and movement of people includes movement of cattle, potentially carrying the infective parasite.

On the heels of political turmoil in the country, including the reign of Amin, full-fledged civil war, and ongoing rebel activity in the north, sleeping sickness control in Uganda has been decentralized since the 1990s. This decentralization, combined with a decline in the number of patients with sleeping sickness resulted in reduced investment in disease control.

Dr. Mbulamberi, now a senior official in the Ministry of Health, is still gravely concerned about sleeping sickness and his personal commitment and emotions regarding the disease are clear during our conversations. The current leader of the sleeping sickness control team, Dr. Abbas Kakembo, is animated and focused. He says that their current work with sleeping sickness is more akin to "putting out fires," with limited attention on prevention. Dr. Mbulamberi's bicycle program eroded as priorities changed. To put it simply, when the bikes broke, the money to fund the program was going to other things. Yet from my own experience and research, the control of the disease does require local and on-the-ground solutions such as Dr. Mbulamberi's.

Sleeping sickness is a disease of the poor; there is extremely low risk to travellers and most cases occur in rural areas where people have limited access to healthcare. Without treatment, mortality is guaranteed, and we believe that many cases go unreported. How do you bring diagnostics to these rural areas? How do you bring treatment to these areas? How do you monitor and detect disease in such areas? And how do you do solve all of thee problems in the face of all the other health challenges of many sub-Saharan African countries including malaria, HIV-AIDS, and tuberculosis? These are the questions and challenges facing public health workers, such as Drs. Kakembo and Mbulamberi.

Sleeping sickness is caused by a trypanosome parasite that is transmitted from infected cattle to other cattle and humans through the bite of a tsetse fly. Strategies have been discussed to eliminate the tsetse fly, which is the vector for sleeping sickness across the African continent, by using the Sterile Insect Technique (SIT). The theory with SIT is to sterilize male tsetse flies using irradiation and then release the sterile male flies to breed. This could then reduce the population to a level that routine fly control mechanisms would eradicate the tsetse fly completely. In order to do this, we would need to be able to sterilize sufficient tsetse flies and release them on a wide scale across Africa. This requires enormous organization and investment and it's an all-or-nothing strategy. If it doesn't work, the surviving flies will rapidly re-invade and re-establish. How do we coordinate such a venture given that many affected countries are politically unstable? And unfortunately, these discussions about eradication have at times diverted focus from local initiatives for sleeping sickness control.

Although sleeping sickness is a preventable disease, any attempt to control (and I use the word 'control' rather than 'eradicate' intentionally here) the disease needs to take into account the limited available resources, both financial and human. With the necessary support and funding, and a renewed and comprehensive focus on local solutions, Uganda's sleeping sickness team has the dedication and expertise to do their job. I, for one, am hoping that they can. The implications are important.

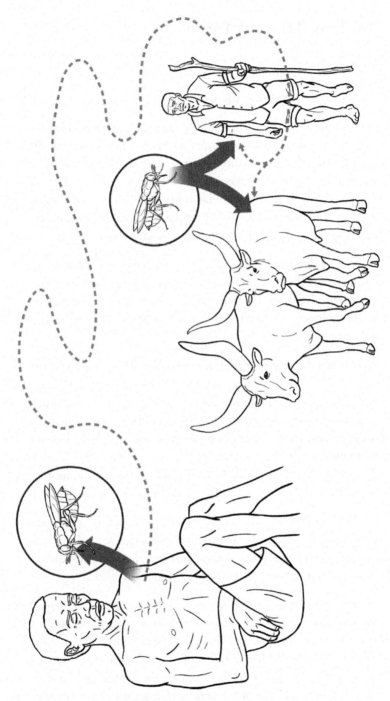

Figure 9 Sleeping Sickness

The Dog Days of India

Sarah Totton

Sept. 19, 2005: Sampling the Goat

At a farm on the outskirts of Jodhpur, at the edge of the Thar Desert
in northwest India, a goat lies under a sack in a farmer's shed. The
farmer reports that it died three hours earlier. For four days, it had
been abnormal – bleating incessantly and staggering. When the owner
put water down for it, the goat had dipped its nose into the water and
then reared its head back without drinking. He also says that two
months ago, the goat had been bitten by a stray dog.

All these symptoms (vocalizing, unsteady gait and being repulsed
by water), plus the dog bite, suggest the goat died of rabies. To
confirm this diagnosis, we must collect a sample of the brain because
the rabies virus tends to concentrate in the brain. This means removing the goat's head.

My colleague veterinarian, Dr. Subhash Kachhawaha, asks the
farmer to dig a grave for the goat. Both Subhash and I have been
immunized against rabies, but this is no guarantee of protection. As
the rabies virus can be transmitted through broken skin, surgical
gloves are mandatory. That is a problem because we were called to this
farm unexpectedly and only have one pair of gloves. The rabies virus
will degrade rapidly in the 30ºC heat. If we go back to get more gloves,
it might be impossible to detect virus in the goat's brain.

Subhash and I decide I should wear the one pair of gloves and
perform the decapitation because I have experience working in a
rabies diagnostic laboratory in Canada. Once I get the gloves on, I
have to move the goat across the field to the gravesite.

I pick up the goat's front legs in one hand and the back legs in the
other. I'm not wearing any protective clothing, so I try to hold its body
away from me. The goat isn't heavy, but rigor mortis makes it like
wielding a large, goat-shaped block of wood. I set it down several
times to catch my breath. In the distance, the farmer's son is shovelling
the sandy soil quickly. Subhash and the farmer watch me wrestle the
goat over the field and by the time I get to the grave, the farmer's son

has finished digging. I set the goat down with its head positioned over the open grave.

"What took you so long?" inquires Subhash.

"Just give me the scalpel," I reply.

Subhash hands over a scalpel blade wrapped in foil.

"Where's the handle?" I ask.

Subhash grins. I saw him buy this scalpel at a streetside store, but either he forgot to buy a handle or the vendor wasn't selling them. Again, we don't have time to drive back to town on Subhash's motorcycle. The goat isn't exactly in mint condition as it is.

Now it's possible to decapitate a goat by making very precise cuts, but if you are doing it with a 1 cm scalpel blade with no handle, you have to be very careful. The farmer and his son are standing by the grave, apparently believing goat decapitation is a spectator sport. I ask them to move back.

A scalpel blade won't cut through bone – I have to cut through the joints between the goat's neck bones near its head. I roll the goat onto its back and attempt to flex its head, but the neck is stiff and doesn't bend the way it's supposed to.

I cut through the goat's windpipe, a thin-walled white tube. I hit bone. I've started too far back and I have to lift the cut trachea away to expose the joint. The scalpel blade is now very slippery and difficult to grip. I'm paranoid I'm going to drop it into the open grave. I have to angle the blade between the bones to cut the ligament, the fibrous tissue holding the bones together. I can't see what I'm cutting, but eventually the bones separate and I can push the goat's head far enough back to expose the foramen magnum—the hole through which the spinal cord leaves the skull.

Subhash hands me a makeshift "brain sampler" (a syringe with the tip cut off). I insert this through the hole to get a sample of the spinal cord and brain. Brain tissue, especially if the animal has been dead for a while, has a consistency similar to toothpaste. I empty the syringe of brain tissue into a glass vial that Subhash holds with his bare hand. I am careful not to let any of the brain tissue touch him. Then, he caps the vial and seals it inside two plastic bags. I push the goat's body into the grave. Subhash instructs the farmer to bury the goat and to go to the health clinic to get a series of rabies shots in case he has been exposed to rabies.

Of course, we won't know whether the goat was rabid until we've processed the sample in the rabies laboratory. But a small hitch here: there is no rabies laboratory in the city of Jodhpur. In fact, there is no rabies laboratory in the entire state of Rajasthan, although we are in the process of setting one up. In the meantime, the brain sample needs to be frozen to preserve it until the lab is finished. Subhash has a locked freezer at the veterinary clinic where he practices, so we drive there on his motorcycle.

Despite the fact it's a preventable disease, rabies is still a major problem in developing countries. Every year in India, more than 20,000 people die of rabies. Most victims contract the disease from the bite of an infected stray dog.

India has large populations of stray dogs. Unlike wild dogs such as the dingo, stray dogs can't live away from humans because they get their food from human sources. Many of India's streets are littered with garbage, which is the major source of food for these dogs.

Some stray dogs are considered "owned" by the neighbourhood – people put out food for them or give shelter to their puppies. However, no one ensures these dogs stay out of traffic or get vaccinated for rabies. No one takes them to a veterinarian if they are sick or injured. Because no one spays or neuters them, they breed as much as possible, which means lots of unwanted puppies.

In the past, cities attempted to control rabies by killing stray dogs - not just sick dogs, but any dogs in an affected area. History has shown this doesn't work. New dogs from surrounding areas simply move in to occupy the areas where dogs have been killed. A population of strays is re-established and rabies emerges again.

Recently, the Blue Cross of India proposed a more humane solution. Stray dogs would be captured from the streets, brought to a shelter for rabies vaccination and spaying or neutering. The dogs would then be released back into their neighbourhood. The idea is that the large population of unvaccinated stray dogs would be replaced with a smaller, rabies-free one. The program was named Animal Birth Control, or "ABC."

In April 2004, the Marwar Animal Protection Trust launched an ABC program in Jodhpur. The Marwar Trust invited me to monitor the impact of the ABC program on the health of stray dogs and the incidence of rabies in Jodhpur. As well, they asked me to assess how the program affected the size of the dog population.

Sept. 23 2005: Counting the dogs

If we were to count every dog in Jodphur, it would take months. Thus, we have strategically chosen several districts to survey. Our plan is to repeat the dog-counting ritual in each of these areas at a later date to see if the number of dogs has changed.

We start counting dogs in Shastri Nagar, a wealthy residential district in Jodhpur. The sandstone houses are palatial and the streets are wide, with little garbage. This is in contrast to the Old City, which lies within the walls at the heart of Jodhpur. It has narrow alleys and open sewers and dogs are plentiful. Stray dogs are not inclined to sit quietly for us, so we use a special technique to count them. Two people work as dog markers, walking through the area before us. Each of them has a plastic canister strapped to his back attached to a hose with a nozzle at the end. The canisters are designed for use against mosquitoes, but the dog markers have filled them with beetroot juice.

They have been spraying the dogs all morning, not always with complete accuracy. We can see where they've been by the magenta stains on doorsteps, lampposts, gates and cars in the neighbourhood.

The counting team, which includes me, Chandra Kant Mishra, our translator, and Dr. Alex Wandeler, a world rabies expert, counts each marked and unmarked dog we see. We plug the number of marked and unmarked dogs into a complex equation to determine the total number of dogs in each area. All dogs that are spayed or neutered through the ABC program are given a notch in their left ear. We can then calculate what percentage of dogs in a given area have been vaccinated and spayed or neutered.

Most of the stray dogs in Jodhpur are tan or gold in colour, short-furred, larger than a border collie but smaller than a retriever, and only rarely do they resemble a recognizable breed. Most of them are thin and while they are not usually aggressive, they do not come close enough to let people touch them. When I assess the dogs' general health, I have to drop biscuits on the ground to entice them to come close enough for a brief visual examination. It is not safe to handle stray dogs on the streets so my examinations are of necessity "hands-off".

At first, not everyone in Jodhpur was convinced what we are doing was useful. Some of the animals were unimpressed also. Once when I was doing health checks on the strays in a poor neighbourhood, I encountered a dog with severe mange. I threw it a biscuit and as it gobbled it up, a stray white bull came up behind it, knocked it with its horns and threw it a couple of metres. Then the bull charged me. As it chased me around the courtyard, a man demanded that I explain why I was looking at dogs when there was so much poverty and poor health in people. I shouted I'd be happy to answer his questions after I got away from this bull. The man seemed oblivious to the bull and kept turning around to face me as I was running. A few people drove the bull off with brooms, and then I explained that vaccinating dogs for rabies keeps people from getting the disease. He seemed reasonably satisfied with this.

October 8th, 2005: The Gaushala

At the guest house where I'm staying, I get a call from Dr. Mahesh Suman, another veterinarian working for the Marwar Animal Protection Trust, telling me that Subhash has found another rabies suspect. This time it's a bull.

Mahesh picks me up on his motorcycle. I am expecting to be taken to a village farm, but instead, after leaving the Jodhpur city limits, Mahesh pulls under the archway of a large sandstone complex that would look exactly like a university…if it weren't for all the cows wandering around inside.

Subhash greets me, telling me this place is a gaushala, or cow sanctuary. There are many gaushalas in India. This one is called the Shree Ram Kumar Panna Lal Gaushala. Stray cows (prevalent in India) that are sick or injured are brought here for treatment. Though the gaushala receives some government funding, it is mainly supported by charitable donations. The facility is enormous, capable of housing 1,200 animals. Apart from his duties spaying and neutering dogs for The Marwar Trust, Subhash is also the chief veterinarian at this gaushala.

Subhash takes me to the isolation pen housing the suspected rabid bull, a large off-white, emaciated animal with two big, upward-curving horns. Subhash points out the bull's elevated tail as a typical sign of rabies. He demonstrates that the bull is also showing signs of aggression. He waves his hand in front of the bull over the gate and the bull lunges forward and rams the bars. Because the bull isn't owned, we don't know how it may have contracted the disease. It is sacrilege in India to euthanize a cow, no matter how sick. If the bull is rabid, it will die soon anyway.

Subhash then takes me to see another stray cow that is suffering from indigestion. Because of the large amount of litter on the streets and the fact that cows eat all kinds of things, including plastic bags, street cows often become sick. The bags can become tangled in the rumen (one of the four chambers of the cow's stomach) and cause a blockage. A rumenotomy must be performed to open the rumen to remove the obstructing objects.

Having performed many rumenotomies, Subhash will be operating on this cow tomorrow and asks me to assist him. The last time (actually the only time) I did a rumenotomy was in my third year of veterinary school and that was on a sheep. Subhash loans me his large animal surgery textbook so that I can refresh my memory.

The next day, I return to the gaushala for the rumenotomy. Mahesh will be assisting as well. Subhash offers to loan me coveralls, but I decide that the surgical scrubs I am wearing are suitable for a rumenotomy. I think Subhash is too polite to tell me I am an idiot.

The textbook warned it might be necessary to cut the contents of the rumen into small pieces before removing them. It might have been more informative if it had said, "It might be necessary to stand with both feet braced on the operating table while you grasp the rumen contents and haul on the end as hard as you can. It might also be necessary to administer more anaesthetic to the patient at this point and make sure she is securely tied to the table so that she doesn't try to get up and walk away."

The rumen contents do not leave the rumen without a fight. Mahesh and Subhash insist on doing the really hard work – pulling yards of plastic bags, rope, burlap sack and other items from the rumen. We are all coated with partially chewed food by the time we finished. An enormous quantity of mucky debris is also piled up on the floor. I wish we had a scale to weigh it – we could sculpt a small child out of it.

Once the rebellious material is out of the rumen, the exciting part begins. Mahesh, who has the longest arms of the three of us, reaches into the reticulum to search for more foreign objects. The reticulum is the chamber of the stomach closest to the ground. For this reason, if a cow swallows heavy objects such as metal, it tends to collect there. These objects are dangerous as they may puncture the wall of the reticulum and penetrate the heart, which lies close by. Subhash tells me sometimes gold jewellery is found inside the reticulum. Custom has it that the chief surgeon has the pick of the pickings, and the other surgeons get what's left.

Unfortunately, this cow doesn't have the Midas touch. There is no gold. Instead, Mahesh pulls out a fair quantity of interesting objects from the streets of Jodhpur: the button from a pair of jeans (the rest of the jeans were presumably part of the mass of fabric we pulled from the rumen), a piece of colourless glass, a one-rupee coin, a handful of small stones, some leather wadded up into little balls, two shiny nails, a few metal staples and a miniature brass key. I ask to keep the key as a souvenir. We then sew up the cow and leave her to the technicians to watch over her recovery.

We are then told the suspected rabid bull has died; we need a brain sample, but a scalpel blade won't work on an 800 kg bull. Subhash says the bull will have to be transported by truck seven kilometres to the government-owned land for disposal of cattle carcasses. We follow the truck on Subhash's motorcycle to the savannah atop the sandstone cliffs above the gaushala. We see a few dogs, which keep their distance, and a large number of vultures. Vultures, like all other birds, cannot contract rabies.

One of the gaushala workers who is immunized against rabies uses an axe to decapitate the bull. Subhash takes a sample from the spinal cord as well as the brain.

May 3rd, 2006: Rabies Diagnostics Laboratory

Nothing moves quickly in Jodhpur, but there has been some progress since my first visit. The municipal animal shelter, which picks up stray dogs in response to complaints of barking and/or biting, is providing reports to us about suspected rabid dogs in their care. In addition, the government veterinary hospital has agreed to report any rabies cases they see so we can sample the affected animals. However, we still do not have a system in place for detecting all of the rabid dogs in the city.

The majority of rabid animals are killed by people defending their families or livestock, and unfortunately, they are not being reported. An education program is under way, but until we have full co-operation in the community, we will not be able to determine the true incidence of animal rabies in Jodhpur.

Still, there is reason to be hopeful. More than a year and a half since collecting the goat and bull brain samples, a rabies diagnostics laboratory has been built at the Marwar Animal Protection Trust facility. It is situated in a special "dark room."

Dr. Alex Wandeler and I are processing the brain from the suspect rabid goat from September, 2005. The sample had been transferred from Subhash's clinic to a freezer in the new laboratory. A tiny amount of brain tissue is transferred onto a glass microscope slide. The brain is then stained with a special fluorescent dye designed to adhere to any rabies virus. If there is no rabies virus present, the stain will wash away completely when we rinse the slide and there will be no fluorescence.

Then comes the moment of truth. The florescent stain is only visible under ultra-violet light. Alex peers at the goat's brain sample through the ultra-violet microscope. The sample looks like a night sky speckled with green fluorescent stars, a positive test. The goat was rabid. We then examine the brain of the bull from the cow gaushala. It too is positive for rabies. We also analyze samples collected over the last 18 months from suspected rabid dogs. Some are positive and some are not. One of these dogs died at the municipal animal shelter and was positive.

We continue the dog population counts which indicate that the stray dog population in Jodhpur is decreasing. The community is beginning to accept the ABC program with people believing it is helping the dogs. Early health checks indicate the spayed or neutered dogs are gaining weight and fewer are emaciated. Things are looking hopeful for the future.

Rabies Control in Rural Kenya

How a Veterinarian Was Able to Follow His Childhood Passions and Help his Community

Philip Kitala and John McDermott

Philip Kitala grew up in a part of Kenya where dogs and rabies were plentiful. The story of pain and hardship was one of many that drove home to him the importance of his research on rabies.

The man was bitten by a suspected rabid dog. The only place he could get post-exposure vaccination was at the district hospital, far from his home. After two weeks getting together money for the bus, he finally got to the hospital only to be told they had no vaccine. They told him to buy some from a pharmacy. He had no money for the vaccine, of course, so he returned home and sold a cow to pay for the medicine. By then the bite wound had healed and he thought his risk of rabies had passed, so he gave the money to his daughter for her school fees. A fatal decision - he developed clinical rabies a week later and died.

Another early experience also influenced Philip. During Grade 5, his headmaster was bitten while riding his bicycle and simply treated the bite with salt and kerosene. About a month later, the headmaster came down with full blown rabies with dramatic symptoms, including barking like the dog that bit him.

In Machakos District in the rural Kenya of his childhood, Philip and his friends had great times hunting, herding animals and exploring the countryside and Philip was always with his dogs. In fact, Philip and his pals witnessed many events involving rabies, such as when the colonial veterinary authorities regularly organized dog vaccination and culling campaigns. The boys didn't understand the reason and procedures for these campaigns. They feared their dogs might be killed. If they heard of a pending campaign, the boys and their dogs would hide for several days in the bush until the campaign ended.

Philip was a good student and eventually studied to become a veterinary surgeon at the University of Nairobi Veterinary Faculty in East Africa. Upon graduation, he joined the veterinary faculty. His early rabies experience had focused his interest and he did research on the diagnosis of rabies and taught the epidemiology of rabies to veterinary students. But he wanted to move beyond diagnostic labs and teaching. He hoped to explore how he could develop and implement ways to control rabies in his home district.

The opportunity to learn new skills in surveillance, epidemiology and infectious disease modelling came as part of a collaboration between the University of Nairobi and the University of Guelph, funded by the Canadian International Development Agency. The program established a veterinary epidemiology and economics training and research program at the University of Nairobi and took on many practical field projects. Under the supervision of professors Joseph Gathuma and Moses Kyule of the University of Nairobi and John McDermott from the University of Guelph in Ontario, Philip developed a project on rabies control for Machakos. He had the enthusiastic support of his colleagues.

Philip wanted a different approach from what he had seen in his youth. He would explore every facet of rabies epidemiology and control. To convince decision makers to do something about rabies in his home area, he had to demonstrate the extent and economic impact of rabies and what could be done to improve a bad situation.

Anecdotes and practical observations abounded, but hard data evidence was lacking. Rabies was well known, but most people were too poor or did not know what to do if bitten. As a result, most people were treated with local remedies and not reported to government officials, and it often led to fatal consequences. Thus, to mobilize support for rabies control, Philip needed an approach for a dog rabies vaccination program based on evidence. He found many willing collaborators among the international community on rabies control. A wide variety of international experts provided him guidance, from how to study dog behaviour and ecology to disease surveillance and molecular diagnosis to epidemiological modelling and economics. Philip mixed this wealth of knowledge with his own experiences, even using some of the lessons and tricks from his youth.

A surveillance system could identify people who had been bitten by a potentially rabid dog. But he would immediately lose credibility in his community if people saw him studying their rabies problem without helping them get treated after a life-threatening dog bite. Philip knew he had to establish a system to provide post-exposure vaccinations and he wanted vaccine to be available in local clinics at the lowest cost possible. He negotiated with clinic staff and the supplier of human rabies vaccine in Kenya to establish this network. Once in place, the program was widely publicized so anyone who may have been exposed knew how to get treatment. This treatment network allowed Philip and his team to establish a good relationship with the local people. The system of post-exposure vaccination of people worked well and still exists. However, even this approach was often insufficient because many people were too poor to afford even the cheaper vaccines. In one instance, three children were bitten. Two were treated and survived. The third, a child of a poor widow who couldn't afford post-exposure vaccine, died.

Even so, with the human vaccination system in place, Philip started to tackle the dog rabies problem. The first task was to make sure his work fit well with on-going government efforts. He met with district and other veterinary officers to discuss his plans and gather feedback, and he visited frequently with these officers throughout the region. He was so enthusiastic and determined that veterinary staff starting at the local levels all the way through to the national diagnostic labs were completely integrated into his program. Once his active rabies surveillance system in dogs was established in Machakos District, it was so effective that the results from this district completely skewed the government rabies reports. Thus, they needed to be reported separately. Other government agencies were engaged to provide consistent messages to communities.

The next task was to estimate the incidence of rabies in people. Philip and his team randomly selected villages and then households for a more detailed follow-up. They recruited a rabies worker from the local community, in agreement with the local chief and elders. These workers would find all cases of rabies and all incidents of human animal-bites in their area over a one-year period.

The community rabies workers were successful, with some going to extraordinary lengths to uncover cases of rabies. Owners of dogs that bit people were responsible for all treatment and associated costs, thus they would sometimes kill their dogs and bury them secretly. When diligent rabies workers learned about this, they would dig up the carcasses and obtain samples for testing. Rabies workers also collected samples from all dead animals, including road kills, for diagnostic testing. About 5% of road kills were positive for rabies.

As part of the information gathering, household surveys on dog ownership and husbandry practices and vaccination histories were obtained. The household surveys usually went well but were not always popular. One time, a man well-known for refusing to vaccinate his large group of dogs, confronted Philip and his team and threatened them with his bow and arrow when they came to his house. However, Philip knew from his youth that schoolboys could find out anything about dog ownership, so he enlisted the help of schoolboys to gather information on reluctant households such as the numbers of dogs, whether they were vaccinated or if they had bitten anyone. This kind of detective work is often necessary for reportable infectious diseases, and led to a complete picture of dog ownership, rabies vaccination and incidence in the study villages.

Schoolboys also conducted a census of dogs in all study villages and asked about previous dog bites and rabies cases over the past year. The schoolboys' results corresponded well in villages with diligent rabies workers, but the program did expose one or two laggards in other villages. As a consequence, the active surveillance system estimated the incidence of rabies at almost one per 1,000 dogs per year, an extremely high rate and 200 times the incidence rate based on the records from the standard case reporting system available at the time.

The logistics of the project required a lot of innovation. The collection, storage and transport of samples for rabies testing were challenges given the limited budget and remoteness of many locations. In study villages, they used small kerosene refrigerators and established a regular collection schedule from the rabies workers. The workers received careful training in sample collection and storage. No system for rabies diagnosis in rural Africa is perfect, so Philip's system probably underestimated the incidence of rabies, but only slightly, as there was agreement between information collected by household surveys, the schoolboy census and the active surveillance by rabies workers. Thus, there was confidence in the accuracy of the information.

Procurement of materials, such as bleeding tubes, needles and diagnostic reagents, was also done on a shoestring budget. These were much cheaper in Canada, so every time John McDermott returned to Kenya from Guelph, he was heavily laden with boxes of supplies. As they packed for the return trip to their home in Kenya, John often had to debate with the other three members of the McDermott family about whether project materials took precedence over books, toys and kitchen supplies!

This basic rabies information gained from the active surveillance program was critical in the comparison of different vaccination and control strategies. Census information revealed the dynamics of the dog population that helped explain why rabies has been so persistent and difficult to control in Machakos District. One feature of dog husbandry that emerged was that dogs could not rely on food from their owners to survive so they foraged widely, increasing their contact rates with other dogs. Dog populations were quite dynamic with half of the dogs less than one year old.

The research team collected other field data to examine the transmissibility of rabies locally (the so called 'basic reproductive rate of rabies'). One way of estimating this from field data was to track dogs that were known to have bitten multiple people and/or animals, and to calculate how many other people or animals they bit.

With all of the data collected, Philip and his colleagues from the London School of Hygiene and Tropical Medicine and the Ontario Veterinary College modelled and assessed different options for rabies vaccination. During his three weeks in London and eight months in Guelph he compared and discussed options for vaccine frequency with experts.

A hotly debated question around the Kenya work was the role of wildlife in rabies transmission. Some major wildlife reserves surround the Machakos District and small wildlife is quite common in the community. Team members undertook a variety of sampling methods for wildlife, including night transect walks and live trapping and release. After the data was analysed using population models for multiple animal species, it was concluded that wildlife did not play an important role in maintaining rabies transmission in dogs. Evidence showed that rabies is maintained and transmitted within the dog population. Dogs can spread the disease into wildlife populations, but rabies isn't maintained in the wildlife species. As a result, dog vaccination programs near wildlife reserves can protect small and susceptible wildlife populations.

In general, persistent rabies transmission from dog to dog in a population requires a dog density of at least five dogs per square kilometre. Notably in Machakos District, the number of dogs ranged from six to 21 dogs per square kilometre in rural areas and 110 dogs per square kilometre in urban areas. These results helped to explain why dog rabies does not persist in more sparsely populated areas of Africa and fit well with the distribution of rabies incidence across the African continent.

Philip's Kenya-focused study had broad applications to other areas of rural Africa. His work as a member of the Southern and Eastern Africa Rabies group has been valuable and allowed countries to standardize the reporting of rabies and the recommendations for control.

While the results of this research showed what had to be done, establishing a sustainable program is complicated. For effective control of rabies in a population of dogs, at least 70% of the dogs needed to be vaccinated once per year . If the vaccination program is conducted twice a year, only 55 to 60% of dogs needed to be vaccinated each time. In addition, the life expectancy for dogs overall was only 2.8 years, and it was much lower for female dogs than males. With such a rapid turnover of the dog population, more frequent vaccination is required because new puppies are always entering the population. Unfortunately, our estimate of the overall vaccination rates in the District, based on the household surveys and serology, was 29 per cent.. This rate of vaccination is reasonably good by developing country standards and reflects knowledge of rabies in the area. Unfortunately, this rate is too low to control rabies in the population.

The problem of rabies continues to increase across Africa and Asia as populations of people and dogs expand in urban and rural areas. Tackling the problem is more an organizational than a technical problem. Most successful models have relied on community participation with support from government veterinary staff because the government veterinary services don't have the financial and human resources to sustain vaccination campaigns. Colleagues in other developing countries frequently ask Philip to help them tackle their own rabies problems. While much has been accomplished, much remains to be done to protect people and animals from this ancient and terrible killer.

Figure 10 Rabies

A Dog's – and Cat's and Rat's – Life:
A Veterinarian's Tale

J. E. Blake Graham

"Between animal and human medicine there is no dividing line–nor should there be. The object is different but the experience obtained constitutes the basis of all medicine." Rudolf Virchow (1821–1902)

I did not know at the young age of seven, as I observed the animals around my family's farm, that it would be my destiny to one day save a man's life. If I had known then what I know now, I wonder if I would have been terrified or inspired to even closer observation of those animals. I also did not know then that the animal world I observed daily would one day give me the opportunity to make a difference in battling emerging diseases and help shed light on zoonoses. In fact, back then, I imagine that I might have guessed that *zoonoses* had something to do with the snouts of animals, not diseases!

In 1947, after a five-year detour, I embarked on the most satisfying part of my life: I entered Ontario Veterinary College (OVC) in Guelph, and found my life's purpose. To say my years at OVC were seminal to my life's success would be an understatement. Little did I know as I started that I would learn about the animal world in such a profound and intimate way that it would far exceed the simple, scientific knowledge of animal anatomy, physiology and pathology and bring me into a lifelong passion and respect for *human* life.

In my day, veterinary students were required to spend two months in a meat packing plant and two months with a rural practitioner. For part of the summer of 1950, I was posted in Stratford, Ontario during a time when an epidemic of bovine tuberculosis infected the cattle of Ontario. *Mycobacterium bovis,* the causative agent of bovine tuberculosis, is easily transmitted to humans and vigilance is required. In the first half of the 20th century, bovine tuberculosis probably caused more deaths of cattle and humans than any other infectious disease. Affected animals had to be slaughtered to arrest the spread of the disease. In one heartbreaking situation, a farmer lost his entire herd of Angus cattle, about 50 altogether. Milk pasteurization (killing harmful microorganisms through heat) is largely responsible for preventing the disease in people and compulsory testing of cattle has largely eliminated the disease in cattle in the western world.

For the rest of that summer, I was posted in Brigden, Ontario and I encountered another infectious disease of cattle that is easily transmissible to humans: brucellosis caused by *Brucella abortus*. This disease, responsible for spontaneous abortion in cattle, may be transmitted to humans from unpasteurized milk from diseased cows, as well as by direct exposure to the bacteria. When a cow aborts, she often retains the placenta in her uterus and it must be removed manually by the veterinarian, or in my case, the veterinary student. At this time, long rubber or plastic gloves to cover one's arms weren't in use, so I inserted my *bare* arm up to my shoulder into the cow's uterus and removed the placenta.

The next spring I became quite ill with fever, rashes and general muscular discomfort. The college physician and Dr. Don Barnum, head of the bacteriology laboratory of the veterinary college, discovered that I had brucellosis, better known as undulant fever in people. High temperatures are often followed by waves of normal body temperature.

Most likely I had contracted the disease from the infected cows I had treated the previous summer. I didn't want to risk further exposure to brucellosis or tuberculosis and subsequent health problems, so I decided not to pursue a career in treating farm animals, opting instead to specialize in the care of companion animals.

One of the most emotionally charged moments occurred a short time after I had started my practice in Toronto. I had employed a young part-time worker, Gladwin, to help on Saturdays. About a year after he started, my wife Barbara, and I left for a brief holiday, leaving the practice in the hands of a replacement or locum veterinarian. Upon our return, the locum informed us that Gladwin was critically ill at the local hospital without a diagnosis of his illness. I immediately telephoned Gladwin's attending physician. As I listened to the doctor enumerate the symptoms – vomiting, severe pain in his lower back, abnormal urine and high fever – I received a flash of intuition. While in California the previous year, I had seen several dogs with similar symptoms caused by a disease called leptospirosis. The disease is often spread in urine of infected dogs and is transmissible to humans. I speculated Gladwin might have mopped up urine in the clinic from a dog carrying the disease and hadn't washed his hands afterwards. The doctor tested for the disease and confirmed my diagnosis. With adequate doses of penicillin and streptomycin, Gladwin made an uneventful recovery. He became somewhat of a medical curiosity in the human hospital because he was their first patient diagnosed with leptospirosis. Without a veterinarian's help, this young man might have died.

Another time, a new client came to the office shortly after his arrival from Malta. He brought his large German shepherd, who exhibited a variety of unusual symptoms including severe weight loss, anemia, cough and eye discharge. I recommended laboratory work, suspecting the disease was serious. The owner declined, stating he did not have time because he was bound for Vancouver. Uncomfortable with his decision but unable to change his mind, I reluctantly started symptomatic treatment on the dog and hoped for the best. A few days later, a Vancouver veterinarian called to ask me about the same dog but again the client did not stand still long enough for proper diagnosis. Within two weeks he was back in my office with a very sick dog. At this point, I referred him to the Ontario Veterinary College because I was now certain the dog had an exotic disease that my practice was not equipped to diagnose. A veterinary parasitologist with extensive experience in tropical diseases diagnosed leishmaniasis, a disease that doesn't occur in Canada. The disease is transmitted by the bite of a sand fly. The disease can be fatal to both animals and humans so, to say the least, all of us who had come in contact with the dog were worried. Fortunately, no transmission had occurred. The dog had obviously contracted the disease in Malta, and unfortunately had to be euthanized because of his highly contagious disease.

The father of a 12-year-old girl paid more attention to my advice. He brought in a skunk for treatment. The skunk was ill and seemed lethargic. Its lower jaw was slack and unresponsive. With the possibility of rabies on my mind, I suggested that the skunk be left in hospital for observation. When it died a few hours later, I removed the head for laboratory examination. I had noticed that the daughter had numerous warts on her hands, some of them open. With the possibility of rabies in the skunk, I told the father to contact their family physician immediately; the child was started on preventative treatment for rabies and never suffered consequences. Later on, the diagnosis of rabies in the skunk was confirmed. A few months later, several groups of men were sitting on the veranda of our golf club when the father of the girl stood up and addressed the crowd. He gave a dissertation on how, in all probability, I had saved the life of his child. I was overwhelmed by his words and felt I had done what most other veterinarians would have done.

Veterinary medicine has grown exponentially since I first crossed the threshold of OVC in 1947. I have always been a keen observer of the animal world – from my very first days on the farm. I have learned many things, not the least of which is how closely the animal and human worlds are tied. It is easy to see, in retrospect, how easily lives are affected: how a dog's urine can mean death through leptospirosis; how a disease such as leishmaniasis could prove fatal to many humans; how rabies could take a young girl's life, if not for quick observation and preventive care.

A few years ago, I made another important decision, initiated by my experience with these diseases - I decided to provide funding for veterinarians to study and do research for MSc and PhD degrees in diseases of animals transmissible to humans. My hope is that this catalyst will continue to make important contributions to the health of both humans and animals.

The Rudolf Virchow quote was included in "Interrelations of Human and Veterinary Medicine," New England Journal of Medicine 258 (4):170-177, 1958.

A Spiral Odyssey

John Prescott

Thursday, 4:30 am: Brindle, the Jack Russell terrier

Don't you hate throwing up?

The gods say "as sick as a dog" but they don't know, as a dog, what it's like to be sick. Am I dying? I can hardly crawl out of bed to throw up. I feel I've been beaten up. My muscles are so painful.

I'm going to get into so much trouble when they come down in the morning. There's sick all over the kitchen. I get so upset when the gods are angry. I feel terrible, almost delirious. I'm burning up, so thirsty. There's no water left. Strange, though I've drunk so much, I haven't peed. I'm going to throw up again. What did I do to deserve this? Am I being punished for drinking from the downstairs toilet? The gods hate that. Was it for not coming back from the park when I met that lovely spaniel?

Maybe it's not my fault. Perhaps someone did this to me. Who? That snotty-nosed Great Dane next door? Or those horrible raccoons that are living in the garage? They terrify me. They piss everywhere. The garage stinks of their crap, but the gods don't even know they're there. I've even heard that raccoons can drown dogs, though I'm more afraid of their teeth and claws. And their arrogance.

Thursday, 5 am: Pomona, the bacteria

It's great to feel quite at home, though everything is still chaotic. Getting from one animal to another is always terrifying because you never know if you're going to make it. It's the waiting and the not knowing that's the worst. But once you're in the kidney, it's usually safe and the food is great. Lots of urea and vitamins and the fats that come from breaking down cells. Once we find the right animal we can live in them for as long as they live.

Why do they give us such stupid names? Leptospira, which means 'fine coil' in Latin. I mean, who speaks Latin these days? And pomona, which is somewhere in California. I mean *Leptospira pomona*, what an insult! Would you like to be called that? And they keep changing our names as they find out more about us. Really, we just want to be left alone and not put in a test tube and studied to death. And I mean, to death!

The only problem about living in the kidney is that sometimes you get peed out. That's when it gets scary. We're really tiny and fragile and we're used to living where it's warm and dark, full of food and where we can tangle up with our family. So suddenly being out in the open is really frightening. You just have to hope there's lots of water around and the temperature is right. If it's freezing or too dry, and if there's too much sunlight, we're goners. Give me a fall day after a good rainfall every time.

Global warming has been really good for us these last years in Canada. A warmer fall means we can spread out more once we're out in the open.

Thursday, 7 am: Brindle

I didn't get into trouble when the gods got up. The main goddess started crying and hugged me. This was very painful because I was shivering with cold and my muscles hurt even more now, but I wagged my tail slightly and whined – and she hugged me even more! My being pathetic brings out the best in the gods. I licked the main goddess's hand weakly, which made her start to sob. They gave me some water, which I soon threw up again!

Thursday, 7:30 am: Pomona

What an adventure! After the raccoon peed me out last week, I swam for a bit in the muddy puddle in the front yard. I could sense others from our pod in the raccoon's kidney had also made it out and survived. I enjoyed the freedom and sense of new beginnings.

What a break! After the dog took a big mouthful of water, I knew I'd found my new host. I grabbed onto some cells in his cheek and started burrowing as if my life depended on it. Which of course it did. I don't know how we learned to dig like this; it comes naturally and we do it well. I started spiraling, just like a corkscrew going into a wine cork, and almost immediately I was inside a cell in his cheek. It's the only way to survive because otherwise I'd be swallowed and end up in a bath of stomach acid. For fragile guys like us, that's game over. Did you know that we can also burrow through skin, especially if it's soft?

Once I'd made it into the cheek, I burrowed into a small vein and then the journey really got exciting. Being washed around the body in the bloodstream is like white water rafting, thrilling but knowing that I will survive even if I capsize. You never know where you'll end up. The brain, the liver, the eye, muscles, the pancreas, it can be anywhere. Once we arrive in an organ, there's so much food, we pig out and divide and divide to produce lots more of ourselves. It's a wonderful party for a few days, where nothing stops us enjoying ourselves and the world seems perfect. But then our host fights back and starts making antibodies – which can kill us! So we head for places where the antibodies don't reach, mostly the tubes in the kidney. And that's where I am now.

Did you see my new host lick the hand of the person he calls "main goddess"? One of my offspring, Pomona Jr., burrowed through her skin - that was so smart, but maybe a bit sneaky.

Thursday, 10 am: Brindle

The chief god and main goddess took me to the veterinarian. I threw up twice in the car and shivered all the way. I still haven't peed though I've drunk a ton of water because I'm so thirsty.

Normally, I hate going to the veterinarian. I hate the smell of so many animals, having that glass thing stuck up my bum, being held tightly while they prod and poke, and getting the needle!

This time I was too sick to care because I'm sure I'm dying. I was only vaguely aware of the vet asking questions, feeling me all over, sticking in that glass thing, and getting a needle in my front leg and then my blood flows into a tube! The veterinarian mumbled something about lab tests, but she thinks my kidneys have shut down and my liver is off because I've also gone a bit yellow. How can I have gone yellow? She said something about acute renal failure? What's wrong with my kidneys? I knew I was dying. I wish I'd gone for more walks, been more friendly with the small gods, gotten to know that lovely Spaniel better, and not slept so much.

Now the veterinarian says she thinks I have 'leptospirosis'. What does that mean? It sounds horrible. No wonder I'm dying and puking and can't pee, and my muscles scream with pain. I'm only a dog, how can I be expected to know what leptospirosis means? The gods talk so fast, it's hard to understand what they say even when I'm well.

They're going to keep me here. I want to die at home, not in this horrible place. Now they put another needle in my leg and it's attached to bag of what looks like water. She says there are 'antibiotics' in the water. The chief god and main goddess cry and pat me, so I give a pathetic thump of my tail and look mournful. I am mournful. I'm mourning my passing, going to the great park in the sky. Don't leave me!

Thursday, 11 am: Pomona

Something's going really wrong.

There's something horrible leaking into the kidney tubes, something really painful. My cell wall is getting weak - I think I'm going to explode. This is worse than antibodies! Are these antibiotics? We never talk about them because they're too scary. We can't fight them. Let me out of here, I'm dying.

Monday, 5 pm: Brindle

I'm home! The gods picked me up from the veterinarian office today. I'm so happy. No more cages or needles. I started feeling better the day after I went to the veterinarian, but still puked for two days. And then I started to pee, and it just poured out because they'd filled me so full of water! My muscles still hurt but nothing like when I was dying. It's so wonderful to be home, I'll be a really good dog whenever the gods are looking! This was awful.

Wednesday, 8 am: Brindle

Now the chief goddess is sick with the "flu", so she isn't going to work today. The main god is worried about her. I like having the gods around in the day; it's like a weekend. I still feel stiff and tired so we've only been going for short walks. But this morning, I missed my walk because she's sick.

Wednesday, 9 am: Pomona Jr.

Hey, I'm here!

The dog didn't work out for my mom; the antibiotics killed her. I think she told you I snuck into the dog's main goddess and now I'm in her liver, which isn't bad. Not as good as the kidney, but we have made a nest here and should survive for months. Eventually we will to get to the kidney and escape through the pee.

Although I don't like the name, I'm proud to be a Leptospira. Although we don't like being outside if it's too dry, too hot or too cold, we're survivors. Give us a new host and we can adapt.

As long as you promise not to tell anyone, I'll let you in on our secret. You'll never guess what it is!

It's not being noticed! As long as our hosts' bodies don't become aware of us, we can survive as long as they live! If they don't know we're there, they don't get angry and try to get rid of us. Sometimes there are just too many of us and then we kill our hosts, which also means curtains for us.

I have many relatives - some of them live in rats, some in horses, some in pigs, some in mice, some in frogs, some in raccoons, some in skunks, some in sheep, some in dogs, some in cows. You get the picture. We're all slightly different but we speak the same language and can understand each other quite well. Of course, adapting to a new animal species isn't easy. No one likes change, and lots of us die before the adaptable ones evolve and then thrive.

Strangely, none of us has ever adapted to survive in people or their cats. I don't know why, but we've never managed. We can get into people, as I've just done, but we don't live very long in their kidneys, so they're a dangerous host for us. Cats are different. We don't seem to be able to live there. Everyone knows that about cats, but no one can figure them out.

Friday, 9 pm: Brindle

Things are getting bad around here.

The chief goddess is still in bed; I haven't had a walk in three days. I'm being neglected; I'm so unhappy. They say if she doesn't get better soon, she'll go to the hospital. How do they expect a dog to hang around, without walks, without a kind word, being ignored, after all I do for them? I spend my life guarding them, looking out for them, thinking they're gods, and they treat me with no respect. What a dog's life I'm having!

Saturday, 11 am: Pomona Jr.

Hey, it's not bad in the liver, and some of us got into the kidney and others into the fluid of the brain. There's none of that terrible antibody around here, so we'll do well.

As I was telling you, we don't usually stick around for long in people. But I've heard it said that we're the most common infection people can catch from animals in the whole world! Doesn't that make you proud of us? Wherever there's water and our animal hosts, we do a really good job getting into people. I've cousins who've made their way into people working in Asian rice paddies, people cutting sugar cane in the West Indies, athletes swimming in a triathlon across lakes in Michigan, farmers milking cows in New Zealand, and trappers catching raccoons. I wish I had time to tell all their stories. I'm so proud of them all.

Tuesday, 10 am: Brindle

The chief goddess has gone yellow! I think she must have leptospirosis just like me, but her doctors think she has a virus. Why do they always think a virus causes everything they don't understand? It's so obvious - she smells just like when I had this. I know it's hard to believe, but sometimes I think the gods must be stupid. They don't seem to be able to smell, to see or to hear well. Right now, they can't figure out the diagnosis when it's staring them in the face.

If I try to attract their attention by barking or whining, they get angry. How can I tell them? Why can't they smell it?

Sunday, 6 pm: Brindle

The chief goddess is getting better. Finally one of the gods figured out the connection with what happened to me. I don't know why it took so long, but everyone is very excited and the whole family is taking antibiotics. They are talking of giving me the needle to stop me getting it again, which seems stupid to me. If you have such a terrible disease, surely you'll never get it again?

Sunday, 10 pm: Pomona Jr.

What a terrible adventure! I got out of her just in time! Now I want to find a proper host and never leave home again. It's all very well being adaptable and adjusting to different hosts, but in my opinion, there's too much left to chance. So many of us never find another host after we're peed out or we're killed by antibodies or antibiotics. Still, I don't regret the adventure - it's the only way we can survive and use all the skills we've perfected over millions of years.

See you again soon!

Figure 11 Leptospirosis

Protecting Soldiers from Inside the Laboratory

Don Barnum

It was far from a straight line between being born and raised in Campbellford, Ontario, and taking part in crucial, ground-breaking research to counter germ warfare during the Second World War. But they were connected, and in the end I gained a fascinating insight into the intricacies of undercover work, spying and code-breaking during wartime.

We were toiling away making botulism antitoxin in our military lab at Queen's University in Kingston, Ontario, half a world away from where much of the hostilities were taking place. What's remarkable is that none of the botulism antitoxin that we developed was ever used – and at the time our work was not revealed.

And, strangely, it was not until years after the war that I finally found out why our work was neither used nor discussed.

One certainty around those challenging days during the war was that the lab work I was doing was a far cry from what I expected when I headed off to the Ontario Veterinary College (OVC) in Guelph in 1937.

Indeed, some of the roots for this story can be traced back to the First World War, when chemical warfare with mustard gas and chlorine began to be used on the battlefield. In some respects, I would say, these set a precedent for subsequent bacterial weapons, such as botulism. And from the standpoint of intelligence gathering in the First World War, the British in particular worked feverishly to intercept German communications. Perhaps the most famous intercept at that time was the message from the German foreign ministry to its ambassador in Mexico that Germany was planning an invasion of the USA, information that helped push America into the Great War.

Of course as I mentioned all of this was far removed from my motivation to study veterinary sciences.

In those days growing up on a farm, it would have been customary for me to remain on the farm after I stopped going to school, however far along I had gone in that schooling. Moreover, we had two farms and the land had been owned for generations of Barnums. Actually, the entire Barnum clan in North America, which is of Barnum and Bailey circus fame, can trace its roots to an individual who moved to Boston from London, England.

But even though most sons did take over the family farm, my parents were really forward-looking individuals, particularly when it came to education. My brother, who was eight years older than me, studied pharmacy and became a pharmacist in Toronto. I was also encouraged to get an education, so unlike many of my friends, I finished high school. But after graduation with our family having the two properties, I became very active on the farms.

That might have been the end of it, but circumstances altered my future. I was busy as a young man with a variety of groups, such as junior farmers, and I had become interested in purebred cattle. Then a neighbour who was active in the community encouraged me toward a new career path, pointing out that I was the only young man in the immediate area who had completed high school. He then added that we lacked sufficient veterinary services, suggesting I should consider veterinary school. With the need for a veterinarian in the area, com-bined with my growing interest in farm animals such as purebred cattle and my enjoyment for learning, I decided to try for veterinary college. I was fortunate to be accepted at the OVC.

The plan was for me to return home as a veterinarian and settle into our second farmhouse. But at OVC I fell in love with academic work and research. Nevertheless, I continued to help out on the farms whenever I could, until health problems forced my father to let the farms go.

As veterinary college students, we would find work in the summers with veterinarians, a sort of apprenticeship. In my first year, I worked for a veterinarian in Campbellford who wasn't a particularly admirable man, but most things in life can still be learning experiences. That fall I returned to school and was quite successful, with marks that were near the top of the class, so that success was propelling me through school. The second summer, I followed a routine similar to the first, working with the veterinary practitioner and helping on our farm.

As I came back for third year of veterinary school, the Second World War had started. The campus at Guelph, which was home to the Macdonald Institute and the Ontario Agricultural College along with the OVC, was turned into a training base and I became more involved with the military. It made for a busy year, during which I also served as class president. At the end of the school year, I was offered a chance to stay and work at the veterinary school for the summer as a student assistant to faculty. Very few students had this opportunity, so I was fortunate. At the time when a number of my schoolmates were leaving to join the armed forces, I was able to hone my research interests.

One of our faculty members was Dr. Andrew McNabb, who had graduated in the early 1920's and gone into laboratory work. Dr. McNabb was the head of diagnostic medical services for the province of Ontario, which operated a series of labs throughout the province to which doctors could send blood, urine and other samples. Even though he was a veterinarian, Dr. McNabb was in charge of these public health labs and he would come to Guelph to teach public health. Fortunately, when I graduated in the spring of 1941, he offered me a job at one of the labs and the dye was cast. I was based in Toronto and the path of my career changed to diagnostic instead of clinical.

Nowadays it would be rare to find a veterinarian in charge of a public health department, however at that time it was recognized that microbiology is basic to both veterinary medicine and human medicine. In the lab you use the same techniques and procedures in identifying the agent of the disease, from diphtheria in humans to influenza in swine, and many diseases are shared by both humans and animals.

Thanks to Dr. McNabb's role, we were exposed to a wide range of work. For example, the department of health ran a number of mental institutions throughout Ontario Many of them were in small towns such as in Woodstock and Orillia and they would have farms with chickens or cattle. Dr. McNabb was in charge of these farms and I also had some responsibility for these farms. We did some ground-breaking work on these hospital farms. For example, we introduced the first artificial inseminations of cattle in Ontario. We had learned this technique from a Russian scientist in our lab who had done some of the early work on artificial insemination in Russia.

During this period we observed a vivid illustration of disease transmission. From our Toronto base, we did a lot of work for military people and there were a lot of outbreaks of disease in the army camps. One Air Force base was at the Canadian National Exhibition grounds in Toronto and people from all Allied nations, such as New Zealand and Australia, were stationed there and illnesses such as colds and pneumonia were constant. One of our researchers was working on these problems and figured out that in the morning when all the recruits made their beds they were in extremely close quarters, causing illnesses to be spread. As a result, administrators decided to hire outside individuals to make the beds, which was popular with the recruits! As it turned out, the visiting soldiers did not have the same health defences against these diseases as did Canadians, thus they were especially vulnerable to illness.

At this time I was married and had one child, but I always felt as a young man that I was certain to end up drafted. Dr. McNabb kept finding ways to keep me working in his labs, but eventually I joined the Canadian Army Medical Corps and left the public health service.

In the military, I was initially in charge of a small lab in Toronto doing primarily hospital work; then I was sent to a lab in the department of bacteriology at Queen's University in Kingston, Ontario. We focused on an antitoxin against botulism, which was expected to be used in germ warfare. The lab was run by the head of microbiology at Queen's and we had a mixture of service people and civilians. I hasten to point out that I was merely part of that team and our collective work involved preparing the antitoxin against botulism, determining how it could be used and trying to understand more of the disease. We had a large team and I would describe my part as small. The work at our Kingston lab and related work are documented in the book, *Deadly Allies: Canada's Secret War, 1937-47*, by John Bryden.

And now we arrive at the irony in the story. In the end, we had produced enough antitoxin to immunize the entire Allied forces, but it was never used! It ended up that our antitoxin for immunization of the troops was not needed. The Allied High Command, using intelligence gathered via the captured Enigma machines and code-named "Ultra," had discovered that the Germans had decided not to "weaponize" botulism. However, the sensitive "Ultra" findings were kept secret, even from the Canadian government. The Allied Command had determined that if they announced that antitoxin production and subsequent immunizations had been halted, it would have tipped off the Germans to the fact that their secret communications had been broken.

At the end of the war I was working in a military hospital in Sussex, New Brunswick. The previous year I had received a letter from Dr. McNabb saying he had a chance to be dean of OVC and he asked if I wanted to return to the department of health or become a professor at OVC. My preference was OVC because I was a veterinarian, I knew microbiology, and I knew that microbiology was an important part of veterinary medicine.

While the development of the antitoxin made for an intriguing story from the war, in subsequent years my colleagues and I worked in a number of important areas, including formulating and introducing control, treatment and diagnosis of mastitis, brucellosis, strangles, enteritis, leptospirosis, salmonella and clostridial infections.

Work in the areas of *E. coli*, salmonella and mastitis stand out for me. I was fortunate to work closely with Dr. Carlton Gyles and others on the role of *E. coli* in pigs and calves and I am proud to say we laid the basis for the present knowledge in this area. When I came back to Guelph after the war it was recognized that certain strains (there are a couple of hundred of them) produced diarrhea in humans and animals. With Dr. Gyles, we were able to identify the different virulent strains. The research and knowledge we developed really helped figure out how the *E. coli* outbreak occurred in Walkerton, Ontario in 2000. The town's water had become contaminated with *E. coli* and seven people died as a result, while 2,500 others in the town of 5,000 became ill. Understandably, people working at the Walkerton crisis went to scientist veterinarians like Dr. Gyles for assistance.

All of my subsequent work was rewarding, of course, although none had the intrigue that accompanied our work on the antitoxin against botulism during WWII.

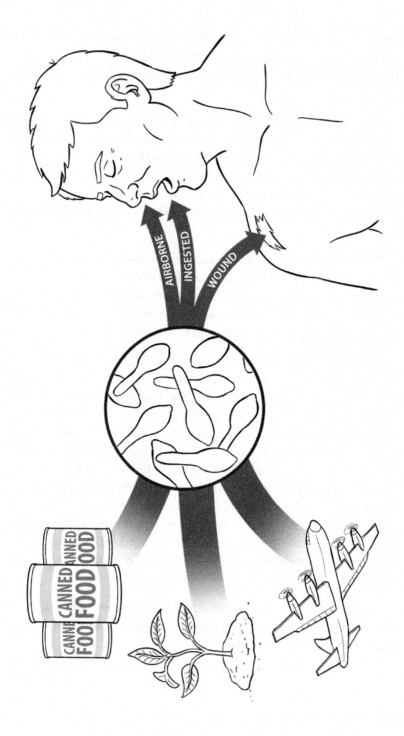

Figure 12 Botulism

Anthrax – A Disease of Man and Beast

Carlton Gyles

Marg stroked her long dark hair with an expensive brush made from imported horsehair. She loved the way it massaged her scalp.

Professor Ingram, a well-known immunologist at the Ontario Veterinary College, pulled on his gloves and conducted a routine postmortem examination on a cow that had died suddenly. There were no specific clues as to the cause of death but he noticed that the cow's spleen was dark and enlarged.

Elsewhere after a long day at the office, Ron went for a jog through the quiet forest and didn't think much of it when an insect bit him on the neck.

And then there was Pauline. She had attended a drumming event in New Hampshire, where she had admired the impressive array of drums with their taut animal hides and decorative markings.

A week after each of these incidents, each of them - Marg, Professor Ingram, Ron and Pauline - had dark lesions surrounded by blisters on their skin. They went right away to their doctors, who took small samples of fluid from the lesion and sent the samples for testing in a laboratory. Given their histories and the results from the lab tests, a diagnosis of anthrax was made quickly. They were treated with penicillin, which led to a recovery.

Chuck and Tony were not as fortunate. Chuck was a farmer who butchered a sick cow and then ate the meat that had not been inspected for disease. Six days later he felt ill, was unusually tired, and developed a fever. He died in a hospital two days later. The diagnosis of anthrax was made from samples taken during his autopsy.

Tony worked in a factory that processed wool and he developed signs of respiratory disease, fever, and nausea. He was diagnosed with anthrax. Despite appropriate antibiotic therapy, he died four days after he became sick.

These rare occurrences of anthrax in humans typically involve individuals who had contact with animals or their products. For example, the horsehair in Marg's hairbrush had contained the bacterium which causes anthrax. Her experience took place a long time ago, and today thorough precautions are taken to ensure that hair and hide imported from countries with frequent outbreaks of anthrax are free of the deadly bacterium.

In the case of Professor Ingram he had nicked himself with his knife during the post-mortem examination of the cow. The cut provided an easy entry point for the bacterium that had killed the cow.

Dr. Ingram's case was well documented and the subject of a scientific paper by Dr. Don Barnum and Dr. Ingram in 1956. During post-mortem examination of a cow that had died very acutely, Ingram observed that the spleen was enlarged and the blood was dark. Microscopic examination of a specially stained smear of the spleen showed that there were capsulated bacteria typical of *Bacillus anthracis*. Confirmation of the identity of the bacterium was obtained from culture and inoculation of mice and guinea pigs. Five days later a small papule, about 5 mm in diameter, appeared on Dr. Ingram's right forearm, just above the glove line. When the papule was opened and the contents smeared, stained and examined, typical *B. anthracis* bacteria were observed in abundance. Dr. Ingram was hospitalized and treated with penicillin and erythromycin. He had enlarged axial lymph nodes, fever, vertigo, headache, diarrhea and anorexia. The lesion on his arm wasn't painful but continued to enlarge to 20 mm in diameter, despite having been shown to no longer contain the bacteria. A photograph of the lesion was taken and has been used extensively at the OVC to illustrate the appearance of the black eschar (dead skin) that is characteristic of cutaneous anthrax in humans. After 12 days in the hospital, Dr. Ingram was discharged, but the eschar remained for 25 more days.

As for Ron, the insect that bit him was carrying anthrax that it had picked up from soil contaminated by the anthrax bacterium. And Pauline had touched a drum covered with an animal hide that had held the anthrax bacterium for years.

Chuck and Tony ended up as anthrax victims. The cow that Chuck had butchered on his farm was infected with anthrax bacteria and when he ate its meat, he contracted anthrax. In the wool processing mill where Tony worked, there were anthrax spores in the air and he inhaled the spores into his lungs. His form of the disease is often referred to as "woolsorter's disease".

I've presented these examples, all of them based on real life stories, to demonstrate the scope of anthrax and how readily anthrax infection could happen in everyday life. Naturally-occurring anthrax can cause outbreaks in numbers of people, such as the one in northern Colombia in 2010 that caused skin lesions in 77 people and at least 2 deaths.

Many countries have developed national vaccination programs that have reduced the incidence of anthrax. However, it still occurs and veterinarians may not have a good index of suspicion for anthrax because they have never seen it. In addition, owners may not seek treatment for sick sheep or goats and thus anthrax, if present, may go unreported and may infect people who eat the meat.

But anthrax is also a disease of terror and biological warfare and has the capacity to affect large numbers of individuals. Anthrax became well known among the general population following the attack on humans through letters handled by the postal service in the United States in the fall of 2001, a few weeks after the September 11 attacks in New York. The bacteria were sent as a fine powder in letters that were mailed to two U.S. senators and to news organizations. Twenty-two persons from five states were infected, 11 with a skin infection and 11 with a lung infection. All of the people with skin infections survived but 5 of the 11 with lung infections died.

Anthrax bacteria can be killed with antibiotics but this treatment is effective only in the early stages of the disease. If treatment is delayed, the poisonous substance (toxin) that is produced by the bacteria continues to damage the tissue even after the bacteria are gone.

The USA cases in 2001 were the result of deliberate spread of *Bacillus anthracis*. Decontamination of the postal system took two years, at a cost of $200 million. Follow-up investigation revealed the bacteria had been derived from USA army sources. The FBI said that a prominent army biodefense research scientist, who worked with the anthrax bacteria, was responsible. When this scientist committed suicide in 2008, it led some to believe that he was the culprit, but others remained skeptical.

The response to the anthrax attack has been substantial. The USA government poured money into preventing and combating bioterrorism, which is the use of biological agents as weapons of terror and destruction. In addition, new and tougher measures were put in place for scientists working with bacteria and viruses that could be used as biological weapons.

In addition to the intentional contagion as with the instance with the mailed anthrax, there have also been accidental cases. In 1979, the release of anthrax spores from a biological warfare facility in Sverdlovsk, Russia, caused an outbreak in humans and animals up to 50 kilometers downwind of the facility. Ninety-four people were infected and at least 64 died. In 2010 there was an outbreak of anthrax among heroin users, first noted in Scotland; 17 people died. There were suspicions that the heroin had been contaminated by bone meal in Afghanistan but that was ruled out. DNA analyses showed that the *Bacillus anthracis* bacteria in the contaminated heroin did not match any of the 4 major strains found in that country. The strain was shown to match a strain found in Turkey.

Anthrax is an even bigger problem in animals than it is in people and Canada has had its fair share. In the summer of 1887, cattle on pasture on the "river flats along the Speed River" in Guelph, Ontario were dying from anthrax. Horses, sheep, and pigs also died. This development was a cause for much concern, resulting in letters to the local newspaper, debates by the local Board of the township, and a cry for help addressed to the provincial Board of Health. Lay personnel referred to the condition as "a mysterious fatality", with animals dying within a few hours of being observed to be ill. Fred J. Chadwick wrote a long letter to the editor of the newspaper detailing the losses suffered by farmers (30-40 animals had died), speculating that the cause may be "some poisonous plant or parasite" or toxic chemical that had found its way into the river. He remarked, "the veterinaries have been puzzled" and suggested that professors at the Agricultural College could be helpful in solving the problem.

Chadwell's letter drew a response from Dr. Frederick Grenside, an 1879 graduate of the Ontario Veterinary College, who taught at the Ontario Agricultural College. Dr. Grenside pointed out that he had examined and conducted post-mortem examinations on many of the animals that had died and that some of the deaths were due to anthrax but others were not. He went on to say that anthrax was well understood, based on extensive studies and experience in Europe – "all its mysteries have been unraveled". He noted that the wastes from Guelph woolen factories, which use some imported wool made their way into the Speed River, which flooded its banks in the spring; the factory wastes were the likely source of the bacterium that causes anthrax. He also remarked that there had been no infections among workers in the factories. Dr. Grenside advised that farmers should dispose of the carcasses by burning or by covering with lime and burying deeply and that they should remove cattle from fields associated with the disease.

Dr. Peter H. Bryce, Secretary to the Board of Health of Ontario, responded to requests that the province become involved. He indicated that Dr. Grenside's report of the clinical facts were clear and he recommended that, as soon as other cases arose, samples of blood from affected animals should be subjected to culture and examination.

In late summer of 1887 Dr. Grenside mailed a sample of blood from an anthrax victim to Dr. Bryce, who arranged for its examination. A microscopic examination of the blood showed that it contained bacteria that had the typical appearance of the bacterium that causes anthrax. Three rabbits, 2 kittens, and 2 guinea pigs were inoculated subcutaneously with a small quantity of the blood. The rabbits showed transient illness followed by recovery and the kittens showed no signs of illness. The guinea pigs died and it was possible to transmit the infection from the dead guinea pigs to other guinea pigs. The laboratory workers were also able to culture the bacteria on the surface of potatoes that had been cut and sterilized as well as on nutrient jelly.

In the following summer, D. McRae, a local manufacturer in Guelph, noted that "anthrax has been bad again". Despite the advice from Dr. Grenside, it was common practice to skin animals that had died and to salvage the hide. The carcass was then buried or removed to be buried in a manure pile, which was later used for fertilizer.

There are few detailed public records of anthrax in animals in Canada for the following 60 years. However, a 1963 report stated that during the past 50 years there had been 163 outbreaks with the majority being in Quebec (83) and Ontario (63). In Ontario in 1952 two mink ranches were reported to have had 11% and 17% of the mink die from anthrax, contracted from contaminated horse meat. In 1957 six zoo animals, including a lion and a cougar, died of anthrax at a zoo in Moose Jaw. A case which occurred in Ontario in July 1996 is remembered by a number of today's Animal Health Laboratory pathologists, including Dr. Tony Van Dreumel who recalls the case very well. A Holstein cow had died without signs of illness having been observed and was subjected to a post-mortem examination at the Kemptville Regional Veterinary Laboratory. Two findings that raised suspicion of anthrax were a massively enlarged spleen and presence of bloody fluid exuding from the nostrils. Culture proved that the death was due to anthrax. An interesting feature of this case was that the farmer had been digging on the farm and that there had been deaths due to anthrax on this farm 65 years earlier.

Anthrax outbreaks occurred in the summer of 1962 in the Northwest Territories and in the summer of 2006 in Saskatchewan that affected bison, cattle, sheep, white-tailed deer, elk, horses, pigs, and goats. Most animals died rapidly without showing signs of illness. Others suffered fever, decreased activity and fluid under the skin. One constant finding on post-mortem examination was that the blood failed to clot.

Anthrax is not contagious from one infected person or animal to another. Most commonly, human infections arise from association with infected animals or contaminated animal products. Anthrax bacterium lives in soil, so it is not surprising that animals infected are usually those that eat grass and other plants in close contact with soil.

The development of vaccines has meant protection for domesticated animals all over the world, but there are still challenges protecting wildlife populations.

A Tapestry of Zoonoses

Jack Cote

As I look over my career as a veterinarian, I realize I witnessed the arrival and treatment of several diseases that spread from cows to people. As I recall these experiences, I see an interesting tapestry of zoonoses. This familiarity puts a face on the perils of zoonotic diseases for me.

To begin, I know both a farmer and a fellow veterinarian who were infected with brucellosis. In the 1960s, I visited a farmer in the hospital where he had been admitted without knowing what had happened to him. His first-calf heifer was aborting and he was helping her because she was in trouble delivering the calf. One of his children had come to the barn to get him because his wife had gone into labour, and he had to take her to hospital. Very soon after his wife gave birth, the farmer fell seriously ill with severe fever spikes, headache and weakness, a condition made worse because they couldn't identify the cause. Once his blood was tested, it came back positive for brucellosis, also called 'undulant fever'. In retrospect, the cow's fluids had contained brucella and the farmer had become infected. He was in terrible shape. If he had been in a weaker condition when he was exposed, the infection could have killed him. Fortunately, once the cause was identified he received the proper treatment and recovered.

Of course, veterinarians are always at risk for infections, and this can happen in ways one may not expect. Indeed, when vaccinating calves for brucellosis, if you are a little careless and inject yourself with the brucella vaccine, you can give yourself undulant fever. (Brucellosis vaccine is unusual in that it is a live vaccine used for a zoonotic disease – accidental injection with most other vaccines would not cause infection). Once when I was working with another veterinarian, he was out doing his vaccination rounds and had one of those careless experiences – I remember the date, June 1, 1952, because I was off from work. He had finished off a syringe full of vaccine, so he put the syringe in his coverall vest pocket with the needle sticking up and went to get more vaccine. Unfortunately, on the way to his car he tripped and the needle pricked his right wrist. In a fairly short period of time, he became quite sick and he got in touch with me to ask if I could come back early from my time off. He knew what was happening to him – he had become infected from the drop of brucellosis vaccine in the needle. His arm swelled to the elbow and he had it in a sling for a month. He couldn't go out to farms for that entire stretch. As I recall, streptomycin was the antibiotic that was used. It might appear a little ironic that he ended up with brucellosis while he was vaccinating calves to prevent the disease, but it was no joke for him as he was out of commission for a significant period.

Brucellosis is now under control, but in the 1950s and '60s brucellosis was common and caused cattle abortions, especially for heifers having their first calf. They would abort at about seven months of pregnancy and veterinarians were often called to assist. Veterinarians and farmers who didn't wear gloves while helping cows have calves could get infected through skin wounds on their arms and hands. Once we had blood test procedures and vaccination policies for brucellosis, it became easier to control. We had a window of opportunity to vaccinate calves between six and nine months of age. Once the vaccination was completed, we were required to make out an individual certificate for each calf. We put a tag in the ear to identify the animal as being vaccinated.

When I was in veterinary school, all veterinary students were tested for brucellosis once a year. As a student, I interned in Illinois and the first thing the veterinarian there did was take a blood sample from me. After I tested negative for brucellosis, I wore a long plastic glove that covered my arm and hand for any obstetrical work. I could still get infected but the glove provided a major element of protection.

After I graduated in 1951, the first zoonotic disease I faced was bovine tuberculosis (TB), a highly contagious disease that attacks the lungs and can be fatal. With bovine TB, the main risk to humans comes from consuming unpasteurized milk. Cattle with TB would become very thin and emaciated, but we rarely saw TB in cattle once testing was mandated. Cattle herds had to be tested for TB, which involved injecting a drop of tuberculin into the fold of skin under the tail of every cow. When infected cows reacted to the test, they were sent to slaughter, where they would be inspected for any lesions of TB.

In Canada, all provinces have now been declared to be free of TB in cattle, but wildlife continues to be a reservoir as suggested by a recent outbreak of TB in bison in northern Alberta.

Salmonella is another zoonotic disease that I dealt with during my long career as a veterinarian. It tends to affect cattle that are debilitated and comes on as a secondary illness and can cause death. The disease can transfer to humans from contact with infected feces of the animal.

I was at a farm one Friday afternoon to see a cow with a retained placenta, fever, lack of appetite and diarrhea. The farmer had a big herd, with cows and newborn calves in one pen. The four boys in the farm family also played in the pen. After examining the cow, I immediately suspected salmonella, told this to the farmer and then added that if any of his sons got sick to get them to the hospital right away. Sure enough, on Sunday one of the boys got sick and the father raced him to hospital. He related to the doctors the story about the cow at home with salmonella. The child was put in isolation and tested and the sample from the boy had the same type of salmonella as the cow did. The farmer recognized the risks and the severity of the illness, and he was fortunate that this was the first time he had had to deal with it. I had a good idea what the cow had from the clinical signs, but it was important that the farmer understood the importance of acting swiftly if any of the children showed signs of having contracted salmonella.

As another example, I recall a Mennonite family and their very sick baby. Earlier they had parked their buggy with their baby in it near a row of cattle while the mother helped with the farm work. To the best of our analysis afterwards, somehow the baby was splashed with feces containing salmonella and became infected. Fortunately, the baby survived, but this story demonstrates that with salmonella, you need to be wary because it is so virulent.

I've also dealt with rabies, another zoonotic disease. This acute, infectious disease attacks the central nervous system and is usually transmitted by the bite of infected animals. One manifestation of rabies is 'furious' rabies, where the animal is hyperactive and excited. You can tell almost immediately what is happening with these cattle. The other manifestation is called 'dumb' rabies, where the cattle quit eating, have a low-grade fever, reduced milk production, are lethargic and drool because they can't swallow. As a veterinarian, your first inclination might be to think that they had something caught in their mouth and you might try to reach into their mouth to check. But if the cow has rabies, you could become infected from the saliva. I was never tempted to do this without first putting on a protective glove. Once a colleague at the Ontario Veterinary College (OVC) had the experience of putting his hand in a cow's mouth that was drooling and then he learned from the farmer that there had been a skunk in his barn about a week earlier. Because skunks can carry the rabies virus in their salivary glands, my colleague went straight to the doctor and got anti-rabies serum. The post-mortem examination revealed that the cow indeed had rabies.

I can also recall a case of furious rabies. We were called in by another veterinarian and took a photographer from the OVC to document the case. There was a cow stumbling around a pond at a farm up near Orangeville, Ontario. The cow would act as if it wanted to charge at us, but then it would stop. It finally fell down near the edge of the pond and soon died. Verification of rabies required removing the head and sending it to federal labs for testing. To no one's surprise, the diagnosis was rabies.

As was common for veterinarians and veterinary students I was vaccinated every five years with the rabies vaccine. When I graduated, rabies was not considered an issue in Southern Ontario, but it was up north. I can remember being in classes and being told we may never see rabies in southern Ontario, but within 10 years rabies was much more of a concern.

I've described these four examples to help readers understand how diseases can be transmitted from cows to people. Certainly things have changed over the years and I have witnessed much of it from a very early age. My father was a practising veterinarian in Guelph from 1926 until 1948 and he joined the faculty at the OVC, where he taught small animal veterinary medicine. I graduated from OVC in 1951 and worked at OVC until I retired in 1986. Fortunately, all the diseases I described are now preventable and brucellosis, tuberculosis and rabies are infrequently seen in Canadian cattle. Salmonella remains a serious problem for both cattle and people. Scientists at OVC and beyond continue to look for methods for prevention and control.

Tuberculosis and the Irish Badger

Wayne Martin

My story completes a genealogical circle that pulled me back to Ireland, the land of my ancestors. But how would a faculty member at the University of Guelph become involved in the control of a disease like cattle tuberculosis in Ireland?

In the late 1980s, Irish scientists asked me to help them and their students enhance their abilities to control diseases of populations. And then in 1990, I met some key advisors for the tuberculosis control program in Ireland during an international epidemiology meeting we hosted in Guelph and Ottawa.

I jumped at the opportunity to work in Ireland. My great-grand-father was born in 1850 in Fermanagh County in Northern Ireland just after the end of the Irish famine. He grew up on a small farm and local church records refer to his father as a farmer and sometimes a labour-er. The farm, which is recorded on 1800 census maps, has a unique beauty, but eking out a living would have been challenging. He immigrated with my great-grandmother to Canada in 1872 and settled on a farm in southern Ontario. His son and grandson (my father) continued that work, so I grew up on a small beef cow-calf farm. It had a major influence on my career.

When I was invited back to Ireland to help with eradication of bovine tuberculosis it became apparent that solutions to their tubercu-losis problem would not be easily found.

Growing up, I remember veterinarians coming to our farm to test our cattle for tuberculosis – luckily always negative. I had an academic knowledge of tuberculosis control from my PhD training at the University of California, Davis where I had worked with American veterinarians trying to rid the USA of the last vestiges of cattle tuberculosis. Despite my limited personal experience, with tuberculosis eradication, I arrived in Dublin prepared to do my best to help my Irish colleagues.

The bacterium responsible for tuberculosis is a member of the genus mycobacterium. The cause of cattle tuberculosis, *Mycobacterium bovis* (*M. bovis*), can survive for long periods in damp moist environments, but it probably would not persist if it wasn't transmitted from animal to animal (including human to human) where it can cause disease and the resultant contamination of the environment.

Human tuberculosis caused by *Mycobacterium tuberculosis* has killed millions of humans throughout history. Because of its chronic debilitating nature, it has been referred to as "consumption." In humans it is largely a disease of the respiratory tract because typically humans are exposed via inhalation. Although the majority of people infected with tuberculosis do not become clinically ill, the disease is difficult to cure and requires protracted periods of therapy with drugs and sometimes, even today, isolation of infected persons.

Cattle tuberculosis also is most commonly spread via respiratory droplets and to a lesser extent by consuming feed that an infected animal has contaminated. In Ireland, the government control of cattle tuberculosis is through a test and slaughter program whereby cattle are tested annually; any "reactors" are removed from the herd and slaughtered. This program began in the 1930s and by 1964 it had greatly reduced the percentage of cattle herds that were infected. This success led many to speculate that eradication of the disease was "at hand." However eradication has proven to be an elusive goal.

Over the last two decades my major thrust has been to develop accurate and complete databases and train graduate students and veterinarians in epidemiological methodology. As epidemiologists, we use data from a large number of animals to infer about how and why disease occurs. We combine statistics and medicine to develop ways of preventing and controlling disease in populations. On a personal note, the academic and personal friends I made on these journeys have enriched my life greatly.

Many productive meetings were held, but I had learned early in my career that disease control is only as effective as the commitment of the key stakeholders. Thus, I also met formally with farmers' representatives and informally with numerous farmers. A number of issues were raised, but the most common question was about the potential role of badgers as a reservoir for *M. bovis*. (A reservoir species is capable of perpetuating the disease agent in the absence of the major species of interest - in this case, cattle).

Badgers are a protected species throughout Ireland and cannot be killed or otherwise disturbed without permission from government officials. However, in 1974 badgers were shown to be infected with *M. bovis*. In 1988, anxiety about *M. bovis* in badgers peaked when two state veterinarians obtained permission to remove badgers from a large area in one county. By 1991, word had begun to spread that the level of tuberculosis in cattle had drastically declined in that county. However, before I arrived in 1991, I had never even seen a badger, so it was time for me to learn about them!

I began by reading a German zoologist's wonderful book about badgers, their habitats and behaviours. Badgers have coarse brown-black hair, white cheeks and a white stripe from their nose to the crest of their neck. Weighing about 10 kilograms, they are powerful diggers with well-muscled shoulders and front legs. The species of badger that lives in Ireland, the European badger (*Meles meles*), lives in extensive underground dens called setts, with a main entrance and usually one or more secondary entrances. Each family collective has a home territory, which they defend from intruders, and the boundaries are often marked by their latrines. Badgers are nocturnal and forage for food at night, carrying fresh straw and hay bedding back to their dens. However, many of their paths become visible to the trained eye, and can be followed back to the sett. Badgers are omnivorous but mainly eat earthworms and grubs that are found in cattle pastures. They also eat berries and cereal grains, which attracts them into cattle barns and feed mangers and brings cattle and badgers into close contact.

Another potential route of transmission of tuberculosis from badgers to cattle is the cattle drinking water. Badgers often mark their water supply to keep out other badgers by urinating in it. Infected badgers often have *M. bovis* in their kidneys and hence contaminate cattle drinking water through urination. The badgers themselves are probably exposed to the bacteria by contaminated feces from infected cattle. The dark and moist badger sett then becomes an ideal environment to transmit and perpetuate the infection among badgers and chronically contaminate the den. Badgers can also infect each other by fighting and the drainage from fight wounds could recontaminate the sett environment.

Early on, I worked with zoologists from University College Dublin on the distribution and density of different species of earthworms. We found that the population of earthworms was larger in the soil of high quality pasture than in lesser quality pastures. Thus, improved cattle pastures increased the food supply and supported increased numbers of badgers. We believed food supply was a main driver of the number of badgers in an area; therefore badger diets came under increased scrutiny. By relating badger density to the soil type and how the land was used, we were able to make initial estimates of the number of badgers in Ireland.

During the late 1990s veterinarians in the Irish Department of Agriculture Food and Forestry (DAFF) were anxious to repeat the apparently successful badger removal program in other places. Thus, four new areas of approximately 100 square kilometres each were selected and divided into a 'proactive' removal area and a 'reactive' removal area. In the proactive area, as many badgers as possible were removed; in the 'reactive' area only badgers close to farms that had tested positive for tuberculosis were removed if it was clear that badgers were the most likely local source of the cattle infection. After four years, by comparing the subsequent cattle tuberculosis rates between the two areas, it was demonstrated that proactive badger removal greatly reduced the risk of cattle tuberculosis.

Once, large scale badger removal started, I wanted to learn as much as possible from the badgers being culled. Thus, we performed post-mortem examinations on half of the culled badgers and discovered that up to 50 per cent of them were infected with *M. bovis*. We also used "typing" to identify the specific strain of *M. bovis*. Based on these studies, we learned that within a county, the strains of *M. bovis* were nearly identical in badgers and cattle, whereas strain types differed from county to county. This supported our thesis that cattle and badgers locally transmitted *M. bovis* to each other. Further detailed analysis indicated that a particular outbreak of cattle tuberculosis wasn't necessarily caused by the nearest infected badger sett, as their strain patterns of bacteria did not always match. Often infected badgers from the same sett had different strains of bacteria, suggesting that their source of infection came from badgers from another sett or from other infected cattle.

Following these studies, the Irish DAFF appointed a new head to their Wildlife Unit, which managed badgers and deer as possible reservoirs for cattle tuberculosis. Recent epidemiological studies indicated that controlling badgers on a farm-by-farm basis control was unlikely to be successful in controlling tuberculosis. Thus, an agreement was struck allowing DAFF staff to remove badgers from larger "black-spot" areas where cattle tuberculosis had persisted at high levels despite intensive efforts to reduce its frequency. During this program, officials removed an average of 5000-7000 badgers, annually, representing up to 5-10 per cent of Ireland's estimated badger population.

Once the 'black spots' were identified using the central data base on cattle tuberculosis and geographic information system (GIS) technology, the team mapped out the badger setts, managed the restraints used to capture badgers, dispatched the captured badgers, and transported the badger carcasses to a central laboratory for post-mortem examination. Subsequently stomach contents were analyzed for dietary studies and the reproductive tracts were collected for further research. These studies demonstrated both seasonal and geographical patterns of the reproductive physiology and dietary behaviour of badgers. The resulting information has proved useful in preventing cattle tuberculosis as well as helping us to understand and manage the badger population.

For example, after evaluating the stomach contents of more than 600 badgers collected in 2005 and 2006, we learned that badgers consumed plant and animals opportunistically throughout the year. Badgers undergo a false hibernation in winter and eat very little. Earthworms and beetles are consumed spring through fall, while insect larvae dominate the diet in spring. When available, the badgers eat cereal grains and berries. Thus, badgers in Ireland proved to be generalist foragers with seasonal food preferences more similar to badgers in Italy and Spain than those in England, where their reported main dietary component was earthworms year-round, with increased cereal grain intake in the summer.

Although reducing the population of badgers decreased the risk of cattle developing tuberculosis, the amount of land under capture cannot be expanded under the current agreement. Thus, the DAFF veterinarians either needed to negotiate a larger capture area or develop an alternative for the removal program. A field trial of badger vaccination has just been completed and appears to show promise for reducing infection levels in badgers. Thus, the current Wildlife Unit planning includes the removal of some badgers for one to two years in six selected counties and then we will establish a vaccination program for the remaining badgers. Proof of efficacy will require that the vaccination program is at least as effective as continued culling at reducing cattle tuberculosis. Our ongoing fieldwork provides much data for research, and an ecologist is now a full member of our team. His work is providing numerous new facts about the ecology of badgers in Ireland; facts that will inform future policy for cattle tuberculosis control and the management of the Irish badger population.

At present there is renewed optimism among the veterinarians and staff in the Irish DAFF. A number of counties have reduced the prevalence of bovine tuberculosis to extremely low levels and if progress can be continued, eradication is achievable. Eliminating bovine tuberculosis would alleviate a huge financial burden on the Irish and EU governments and farmers, as well as reducing the risk of cattle tuberculosis in humans (the greatest risk of the latter occurs from drinking unpasteurized milk). If badgers can be protected by an effective vaccine, re-seeding of badger populations with *M. bovis* from cattle will be reduced or eliminated. Indeed, perhaps the day will come when Irish badgers also can live free of tuberculosis—we may even call them "goodgers" (with apologies for the pun on "bad").

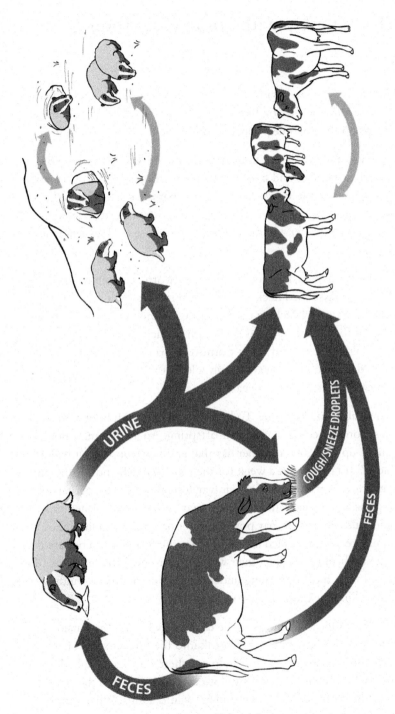

Figure 13 Bovine Tuberculosis

Hooking Up with Some Helminths

Jody Gookin

You know you are having a bad day when you step in a pile of dog poop, even worse if you happen to be barefoot. But, I'd rather step in dog poop than people poop, and not just because it's not as gross!

Lots of parasites live in the intestines (the 'bowels') where they raise their families and boot out their offspring with the poop so the young ones can find new homes. Many parasites are quite particular about where they live, so parasites whose offspring are passed in human feces prefer a human home, while those of dogs will fancy a dog home.

But some parasites, such as the hookworm, can improvise. Named for its crook-shaped neck, a hookworm is about half an inch long and makes a living by sucking on the insides of the intestines. With two pairs of formidable teeth, it latches on in such a way that it mostly sucks blood. Just 10 of these worms can drink a thimble-full of blood every day. And by the way, usually a lot more than 10 are in residence and can live as long as 15 years.

Would you like some? To become infected with hookworms, you must come in contact with their offspring. An adult female hookworm can lay up to 30,000 eggs per day that are then pooped out by her "host." It takes about a week for the eggs to hatch and for the tiny immature worms to wiggle their way to higher ground, perhaps a blade of grass, and lie in wait. So while it may be less gross to step in an older pile of poop, it may pose a greater danger. As soon as the immature worms (called larvae) contact the foot, they penetrate the skin and begin tunnelling their way between the different layers of cells. Like a mole tunnelling under your lawn, they leave winding trails in the skin that cause intense itching.

When they leave the skin, the larvae crawl along the outside of veins to the lungs, where they break into the tiny air sacs. Then they travel up the airways to the windpipe where they are coughed up into the back of the mouth and swallowed. Then, they find their home sweet home inside the bowel and grow into adults. What a trip!

All of that blood sucking comes at a price to the host – and that price is called anemia. Anemia happens when the worms suck red blood cells faster than your body can make new ones. As it turns out, human hookworms travel readily on their trip from the skin to the bowel of a human. Similarly, dog hookworms are competent travelers from the dog's skin to the intestine. However, if dog hookworm larvae encounter a human, they wander around aimlessly in the skin and eventually die because few of them are capable of making it to the human's intestine.

Hookworms were first discovered more than a century ago in miners and rail-tunnel workers, who commonly were anemic and had diarrhea. In thinking about where these men worked, it makes sense why they would get hookworms – particularly if you consider how difficult it would be in a mine or railroad tunnel to find a private place to defecate, not to mention that those people rarely had a decent pair of shoes. Indeed, in the early 1900s about 40 per cent of the population of the southern USA was infected with hookworms. Hookworm offspring love the humid and moist climate of the South and many people defecated in a field or behind a bush. A public health campaign, started in 1909 to build outhouses and encourage people to wear shoes, is credited with nearly eradicating hookworm disease from the South. (If you want to see a video made by the Rockefeller Foundation as part of this campaign, check out the Rockefeller Archives on YouTube, "Unhooking the Hookworm.")

Today, hookworms infect more than 600 million people, most of who are living in areas stricken by poverty and poor sanitation. If you live in the USA or Canada and have a hookworm infection, chances are you picked them up somewhere else. And yet there are some people who want to get hookworms. Why? You probably find this surprising, but here goes.

More than a few well-educated health professionals think urbanized society has become too clean and that a little dirt and disease now would keep people's immune systems from "overreacting" to everyday things, such as the air we breathe and the food we eat. For example, diseases such as asthma and inflammatory bowel disease are nearly epidemic in the USA, but virtually unheard of in countries where sanitation facilities remain so poor that people may still poop behind a bush.

What do those impoverished people have going for them? Worms! – but, of course, it is not that simple.

The scientific thinking is that some hookworms, but not too many, may distract the immune system from focusing on unimportant details such as the presence of pollen in the air or gluten in your lunch. Brave individuals, willing to test this hypothesis, have purposely infected themselves with hookworms and swear by them. While bravery certainly played its part, these folks are also desperate: If you struggle to breathe because of asthma or are driven to the bathroom by painful diarrhea from inflammatory bowel disease dozens of times a day, a dose of worms doesn't sound so bad.

So it turns out that hookworms may be good candidates for a therapeutic worm infection. Just a few worms are not terribly harmful, and the worms can't increase in number inside you as long as you carefully flush their offspring down the toilet. Besides, if you change your mind you can treat yourself with a de-wormer.

As suggested earlier, the problem with hookworms nowadays would be in finding some. Sure, plenty of dogs in the USA and Canada have hookworms and dog poop is not hard to find, but you really need human hookworms for this treatment. Only human hookworms have the predictable talent to travel to the human intestine.

The other major problem is that this whole unorthodox hookworm idea is far off the medical mainstream. Most North American doctors have never even heard of hookworms, let alone be willing to purposely infect patients with them. In fact, it would be illegal for them to do so, so where would you go to get hookworms? Here's the solution: You travel to equatorial West Africa and scope out where the locals duck out to poop; then, you walk barefoot in the human excrement. I am not kidding; there is one well-documented account of someone doing just that.

A slightly less hazardous option is to travel to another country where you can purchase infective hookworm offspring that were lovingly harvested from some human entrepreneur's poop. In this case, the hookworms are fixed to the gauze of a band-aid that you place on your arm. You know you are off to a good start when the skin under the band-aid really starts to itch.

Go ahead – but just keep in mind that poop is like a box of choco-lates. You never know what you're going to get – it may be hook-worms, but it could also be hepatitis, HIV, cholera, typhoid fever and elephantiasis. So, don't believe everything you read!

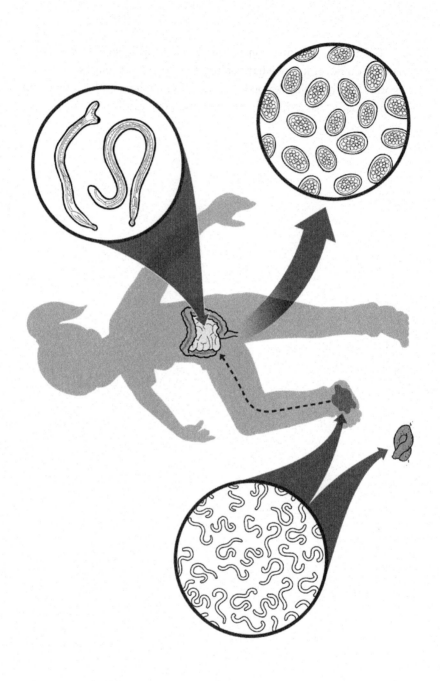

Figure 14 Hookworms

Landing on the Moon

Michele Guerin

It was like landing on the moon. That was my first thought as land became visible from my window seat on the plane. My journey had begun in Toronto. My excitement grew when I arrived in Boston and met my colleague, Cliff, who had flown in from Ottawa. But excitement quickly turned into disappointment when we found that not a single pub in the airport was televising our beloved hockey finals – the Stanley Cup! Even so, the cold beer and lively conversation made the hours pass quickly until it was time for the next leg of the journey. The cabin was dark and I managed to sleep for a few hours on our flight over the Atlantic. When I awoke the sun was up, and all around were fluffy, white clouds and a vast ocean. Then I spotted coastline, but it was like nothing I had ever seen before. The land below was brown and bumpy, and there were no signs of civilization – anywhere! It was like a deserted planet, and I imagined for a moment that I was the first astronaut landing on the moon.

We had not driven very far from the airport in our 4x4 rental vehicle when civilization disappeared again and the brown, bumpy land was all that could be seen. Our first stop was the Blue Lagoon, a massive, outdoor spa with milky-white water that smelled of rotten eggs. I balked at the cost. Cliff assured me that it was well worth it, so I spent my lunch money on admission. The pounding hot waterfall felt like drops from heaven on my shoulders as it melted the kinks out of my sore muscles. As I floated in the steaming-hot, mineral-rich geothermal water, my jet-lagged body began to relax and I started to appreciate the wisdom of my well-travelled colleague. I was too tired to appreciate the beauty of this strange place, but all would be revealed in time.

The next day was down to business. We had come to Iceland to collaborate on a research project. We wanted to learn more about the elusive campylobacter, one of the most common bacterial causes of acute intestinal illness in humans around the world. A person can become infected by eating undercooked chicken or pork, drinking contaminated water or raw milk, or from close contact with infected puppies, kittens or farm animals. Iceland had recently experienced an unusually high number of cases, so scientists from Canada, the USA, Sweden and Iceland were working together to find out why.

Symptoms of campylobacteriosis (the name of the disease caused by campylobacter) begin 2 to 5 days after exposure. I had my own personal experience with campylobacteriosis. Although I can't be certain how I got campylobacter, thinking back on it, I do recall that I didn't wash my hands between handling raw chicken and making a salad just a few days before becoming ill. I remember the experience like it was yesterday. I had every symptom in the book. I had come home from school one day feeling a bit nauseous. Intense abdominal pain and cramps soon followed. By evening I had broken into a sweat with fever. What I remember most vividly was the diarrhea (bloody, in my case) that plagued my sleep, and although not as common, I also experienced vomiting. I became severely dehydrated and exhausted. I felt like I had been hit by a truck! It also made me mad. Usually, the illness lasts about a week. In my case, the diarrhea persisted for three long weeks, and because of that, I missed three weeks of hockey- my favourite thing in the world - including the end-of-season tournament. I guess it could have been worse. A small percentage of people develop reactive arthritis after infection, or more severe consequences, such as Guillain-Barré syndrome, a nervous system disorder that results in paralysis or even death! I was sympathetic toward anyone who became sick from this bacteria and I was highly motivated to help in any way I could.

We met Carl at the agricultural centre in Reykjavík (pronounced Rake-ya-vick), the capital of Iceland, a beautiful, vibrant city on the southwestern coast. When we arrived at the centre, Carl introduced everyone. The language was peculiar but interesting, and the atmosphere was friendly and relaxed. English was spoken widely in the business setting, and for the most part, language was not a barrier. It seemed that our new colleagues were keen to practice their English on Cliff and me.

In early afternoon, Cliff, Carl and I noticed a group of people sitting at the side of the building with their pant legs and shirtsleeves rolled up. Carl went out and asked them what they were doing. Joanna, a beautiful woman with blond hair and extraordinary blue eyes, said, "Are you kidding? Haven't you been outside yet today? Look around!"

It turns out that Iceland was experiencing one of its rare heat waves, and no one was prepared to miss out on it! I chuckled to myself as I thought that it seemed more like a warm spring day than a heat wave. What must winter be like?

By 4:30 pm, I had learned my first Icelandic word – út (meaning *out*). It was Carl's funny way of telling us to leave our Canadian ways (that is, working long hours), at home. I wasn't about to argue, as I was eager to explore this interesting place. As I left the centre, I was thrilled to see several striking Icelandic horses at the pasture fence. They were small in stature and at first glance looked like ponies. But Carl quickly set me straight – these beautiful creatures were most definitely horses. The Icelandic people were very proud of their horse's bloodlines and considered it an insult to refer to them as ponies. Over the next two weeks, I grew very fond of these horses, as I climbed the fence often and fed carrots to my new friends near a thick patch of colourful Arctic lupins.

Sleeping the first few nights proved very difficult. It was early June, and the sun set about 2:00 am and rose again at 4:00 am. Even then, it never really got dark – just dim, like dusk that never quite ends. The light-blocking curtains and eye mask did not help. The sounds of an active nightlife drew me out of bed, and I found myself walking around Reykjavík hoping the exercise would tire me out. The sun was so bright that I had to squint even with sunglasses on. It was on my midnight stroll that I discovered Lake Tjornin, a small lake in the centre of the city. There I sat, in the middle of the night (or day as it would seem), watching the people and the amazing birdlife. I was drawn by the varied species of ducks, swans and other birds, most of which I had never seen before. I went back many times to watch the arctic birds in this peaceful inner-city oasis.

There is a general consensus among scientists that poultry meat is a major source of campylobacter to humans. Unlike humans, the bacterium doesn't make the chickens sick. But where do the chickens get the campylobacter? This was the research question that brought me to this new world. Many researchers believe that the most likely source of campylobacter to chicken flocks is from the environment. Iceland was the perfect place to study this. The poultry industry in Iceland is closed, meaning that there is no import or export of poultry meat; all of the chicken that is raised in Iceland is consumed there. My task was simple - to learn about the bird's environment, such as how they are housed and managed, so that I could help answer this question.

Data collection had already begun before my involvement in the project. Our Icelandic colleagues had started collecting caeca (equivalent to the human appendix) at the slaughter plants from every flock. The caeca is the location in the intestine of chickens where campylobacter lives. They also recorded information on various factors that might influence whether a flock becomes colonized with campylobacter, such as the age of the birds, the season in which they were raised, how the barn was cleaned and disinfected, and the presence of cattle or sheep on the farm, among others. What we were missing was information on the ventilation system on the barns (we hypothesized that wild birds might be responsible for spreading campylobacter to the chickens) and the set-up of the barn floor and roof drainage systems (because campylobacter are known to be able to survive in water). This is where I came in. I needed to actually see a variety of farms to understand the range of barn designs. Once I had that information, I would be able to develop a questionnaire and collect the missing data.

Carl arranged daily farm visits to help us with our research. For me, this was the best part of the trip. How lucky was I to get to travel around the countryside and see this great place while getting to do such important work? I remember one farm in particular. It was the first one we went to and the owner was one of our Icelandic colleagues. The farm was set far back from the highway. From the driveway we could see the ocean as a backdrop to a neat row of barns that were connected by a paved laneway. There were shore birds nearby, but otherwise, it was very secluded and very clean. After walking around the barns, we started to focus in on the ventilation and drains. There were wild birds perched on the roof vents. Not that I blame them - the air coming out of poultry barns is very warm. There was one thing that caught our attention though - the roof vents did not have covers. We became excited that we were on the right track. Wild birds can carry campylobacter. So, if their feces can drop into the barn through uncovered roof vents, it might explain from where the chickens get the campylobacter. Clearly, this was an area we wanted to explore further.

Each farm was a bit different, but all had one thing in common – the warm hospitality of the farmers. After walking around the farm, looking at everything from the manure pile, to the puddles around the barn, to the wild birds perched on the roof vents, we were always invited into the farmhouse for a cup of coffee. The coffee was some of the best I've ever had, which seemed strange considering I was about as far away as one could be from the coffee-growing regions of the world. The only problem was that I didn't speak Icelandic, which was the language most farmers used. Even so, I listened intently as Carl translated for us the lively conversation about current events. As I left each farm, I proudly said "takk fyrir" (meaning *thanks*), and in return, each farmer said "bless". It was only at the end of my trip that I learned that bless was not only a way of saying goodbye, it was also a blessing. I truly did feel blessed!

It was the weekend now, and time to explore this fascinating place. Cliff and I embarked on a journey to Snæfellsness peninsula, a 2-hour drive from Reykjavík on the west coast of Iceland. Snæfellsness is best known for the glacier-capped volcano "Snæfellsjokull", which is thought to be the starting point of the voyage in Jules Verne's classic novel *Journey to the Centre of the Earth*.

Our first stop was a small fishing village. We spotted a few cute houses – you could literally count them on one hand. I had never seen a town so small. When we drove past the port, there were many boats – more than twice as many as there were houses, but no people. We walked along the edge of a cliff until we found a trail that led down to the water below. We spent considerable time exploring the shore and small caves. Unfortunately, it was on our free weekend when we experienced "real" Icelandic weather. The sky grew dark. Darker it seemed even than night. Rain clouds were looming for as far as the eye could see. And then it hit. Strong gusts of wind blew sheets of cold rain sideways across our faces. We looked for shelter. Cliff spotted a house nestled into the edge of the rock face above. We trekked back up and were pleasantly surprised to discover that this house was a small café. It seemed odd that there was a restaurant here considering how tiny the village was. Yet it was here where we found all the people! The homemade lobster chowder, fresh baked bread and hot chocolate warmed me and although I was again stunned by the cost (about $30 Canadian), it was worth every penny.

Our next stop was a black pebble beach, which was unique and impressive with its smooth black pebbles, large boulders and lava rock formations. The pounding waves, vastness and seclusion reminded me of areas of the Newfoundland coast. Reminiscing about my childhood and our coast-to-coast family vacations, I developed a sudden desire to go swimming. After all, I was Canadian – I could handle it. Cliff thought I was crazy, and in no uncertain terms was he going to join me. I did not have my swimsuit, so I took off my shoes and rolled up my pant legs and waded up to my knees. The icy cold water was both shocking and painful. My legs became numb instantly. But, I persevered for about 5 minutes, proud of myself for "swimming" in the North Atlantic Ocean. To this day, I wonder whether I would have been brave enough to take the plunge had I had my swimsuit with me.

The weather did not improve at all. In addition to the sheets of rain, the fog was as thick as pea soup. We drove up the one-lane road to the top of Snæfellsjokull. The drive was scary at best. The road was winding with a sheer drop just beyond the narrow shoulder. I remember praying that we would not drive off the edge or crash into an oncoming vehicle in the fog. I reminded Cliff about where I had put my out-of-country medical insurance, just in case. Cliff glanced at me but said nothing. Either he had nerves of steel or he was crazy in his own way!

At the top, we got out of the vehicle. The ground crunched beneath our feet as we walked on the glacier. We wandered around for a short while but it was difficult to appreciate the surrounding landscape because of the fog. I stood for a moment straining to see anything. I spotted some rock formations and went over to have a closer look. When I turned around, I realized that Cliff had not followed me. He was nowhere to be seen. In fact, I couldn't see anything. I was lost and alone at the "entrance to the centre of the earth". But where exactly was the entrance? What if I accidently fell in? My imagination ran wild. I called out and was relieved to find that Cliff was only a few meters away. We decided to leave Snæfellsjokull because of the fog. But where was our vehicle? Again my imagination got the best of me and I wished we had been smart enough to leave a trail of breadcrumbs to follow. It took a good 15 minutes to find the vehicle in the fog, even though it was nearby.

The drive around the peninsula was amazing. By this time, the fog had lifted somewhat. We could see the ocean on one side and the volcano on the other, and there were waterfalls everywhere we looked. It was late in the afternoon when we stopped in Stykkishólmur, a relatively large town of about 1,100 people on the north coast of the peninsula. It is a centre for fishing and tourism. The town was picturesque with many quaint houses and buildings. We wandered down by the harbour. Unfortunately, there were no more whale watching tours going out that day. I had heard that whales were almost always spotted in this area and I was highly disappointed as this was one my "must do" activities. Oh well - maybe next time. We found a restaurant and enjoyed the "catch of the day". I found it a bit ironic that neither Cliff nor I had eaten anything but fish since we arrived, considering we were working so closely with the poultry industry. It wasn't that we were afraid to eat chicken. If the meat is handled properly (such as using a separate cutting board and knife for preparing raw chicken) and cooked thoroughly we were unlikely to get campylobacter from eating chicken. It was simply because the fresh fish was so delicious.

I continued to admire the scenery on our drive back to Reykjavík. I gazed at the lush pasture on either side of the highway, which gave way to gorgeous mountains. The land was spotted with sheep and horses. The fields were a rich shade of green that is hard to describe, but that reminded me of leprechauns and mythical places. I was pleased that the rain had finally stopped, although I was soaked to the skin and chilled to the bone. Almost at the same time, Cliff and I spotted a double rainbow stretching over the highway. Given the scenic backdrop, darkened sky, and complete lack of civilization, it was one of the most beautiful sights I had ever seen. It was then that Cliff explained something I had been wondering about since we arrived. The brown, bumpy land that I first saw from the plane was actually volcanic rock covered in moss. Over time, as the rock compressed and the moss was replaced with grass, this "moon rock" became the rich pasture that the millions of sheep and Icelandic horses grazed on.

On my flight back home, I dreamt of my travels in Iceland and wondered when I would have a chance to return. I also thought about what lay ahead, as my work on campylobacter was just beginning...

In time, we discovered some of the answers we were searching for. Our preliminary analysis did not support our hypothesis that flocks raised in barns with uncovered roof vents had a higher chance of becoming colonized with campylobacter. This put a damper on our wild bird theory. But for some reason, we could not put the vents out of our minds. It was quite clear that the chickens became colonized with campylobacter mainly during the summer season. This gave us an idea. What if, instead of wild bird droppings getting into the barn through the air outlets, it was flies getting into the barn through the air inlets? After all, some researchers have shown that houseflies are capable of transmitting campylobacter. After doing much reading on the lifestyle of flies (more than I ever imagined I would read in my lifetime), I found the information I was looking for. Houseflies are commonly found in buildings where animals live and they eat just about anything liquid or easily liquefied. At night, houseflies usually rest 1.5 to 4.5 metres off the ground near their food source - which is exactly the same height as ventilation inlets on broiler barns. It seemed logical that the flies might either fly into the barn through the inlets for food, or perhaps even get sucked into the barn accidentally. Other researchers have shown that the biological activity and reproduction of flies is greatly affected by even small changes in temperature. So we took it one step further. We looked at the association between temperatures important in fly biology and the likelihood that a flock would get campylobacter. What we found was that under certain ambient temperature conditions (sustained high temperatures with minimal cool days) fly activity likely plays a role in spreading the bacteria. This meant that we not only knew when to take extra precautions on the farms but it gave us more confidence that keeping flies out of the barn was an important piece of the puzzle.

Since our trip, the poultry industry in Iceland has experimented with putting fly-netting over the ventilation inlets on the barns during the "high risk" summer period, with promising results. We were also able to suggest a few other simple changes that have since helped reduce the number of flocks with campylobacter during the summer season, such as shipping the birds at a younger age, avoiding the use of disinfectant boot dips before entering the barn (if used improperly, boot dips likely make the problem worse), and providing the birds with disinfected drinking water (because the bacteria survives well in water). Of course, as with any research, there were still some questions left unanswered, such as why did cleaning the barns with geothermal water increase the chance that a flock became colonized? We suspect the high temperature water might simply be an indicator for differences in the environmental sources and amount of campylobacter between farms, and even now, further research is underway to answer this question... all because of a trip to the "moon".

E. coli O157 in Cattle and Other Critters

Jan M. Sargeant

American Midwesterners take their college football seriously and it took an experience with *E. coli* for this Canadian to find out just how much!

In 1997 I was at Kansas State University studying *E. coli* O157 in beef cattle. These bacteria make people sick after they consume contaminated food or water. In most people, the resulting diarrhea and abdominal pain lasts for several days, but the illness can be life threatening for children, the elderly and people with compromised immune systems. The prevailing belief at the time was that cattle were the reservoir, where *E. coli* O157 survives. After reading a report that linked *E. coli* O157 in a person to a deer, we were curious about the possibility of deer as a reservoir. Coincidentally, one of our team members, who collected manure samples from cattle for testing, noted that he commonly saw deer in cattle pastures and suggested we also test deer feces while we were on the ranches.

We collected about 200 samples of deer feces over the next few months. The sampling crewmembers walked the edges of the pastures and picked up any deer feces they saw. In this informal study, we couldn't tell when or how many deer had defecated. Even with our relatively poor diagnostic tools, about 2% of the samples tested positive for *E. coli* O157. This discovery grabbed our attention!

Fortunately Dr. David Renter joined our research team, after having just finished his DVM degree at Kansas State. He became my first PhD student in epidemiology and stepped in to continue the work on the possible role of wildlife as a reservoir of *E. coli* O157. For his first project, he devised a novel method to collect large numbers of samples from deer. A native of Nebraska, the state north of Kansas, Dave knew that Nebraska requires deer hunters to bring their killed deer to designated check-in stations. Armed with this knowledge, he sent letters to licensed deer hunters in southeast Nebraska and asked them to bring a sample of deer feces to the check-in stations. He also sent them gloves, bags and prepaid courier boxes for their 'work'!

One weekend, I headed up to Nebraska in a Kansas State University car, complete with logo on the side, to assist Dave. The rifle-hunting season is quite short, so we all pitched in to help collect samples from the check-in stations. I hadn't paid much attention to the football schedule – big mistake! College football is huge in those parts and the Nebraska Cornhuskers, the archrivals of the Kansas State Wildcats were playing Kansas State at home. Kansas State had lost to Nebraska every year for the past 29 years.

Wouldn't you know it? While I was driving around rural Nebraska in my Kansas State car, Kansas State beat Nebraska 40-30! To make matters worse, in a key play at the end of the game, a Kansas State linebacker grabbed the facemask of the Nebraska quarterback and sacked him. The no-call by the referees infuriated the Nebraska fans. And there I was – the visible representative for the entire university – and I don't even understand football! As we opened the bags of deer feces in the security of our lab the next day, we found one bag with a newspaper clipping of the infamous "face-mask play." No note and no deer feces. Just the clipping!

As it turned out, our adventure into "enemy" territory proved fruitful and the hunters sent in 1,600 fecal samples. And we learned that a small number of apparently healthy deer were shedding *E. coli* O157 in their feces.

I started my work on *E. coli* O157 at Kansas State in 1997, which was fairly early on in terms of *E. coli* O157 research. The bacteria had only been recognized as a cause of human illness in the early 1980s. Our ability to identify *E. coli* O157 in those early years was not particularly good and our estimates of its prevalence were probably lower than actual. We did find that in some herds, a high percentage of cattle shed bacteria on some days and then would be negative on a subsequent visit. Even with today's better diagnostic tests, this observation has been borne out: herds appear to shed the bacteria intermittently, with some periods of pronounced shedding.

The cattle don't get sick; in fact, humans appear to be the only species in which *E. coli* O157 causes significant illness, usually as individual cases. However, large outbreaks of disease can occur such as the one in the year 2000 in Walkerton, Ontario, where an estimated 2,000 people became ill from contaminated water and seven died.

My work on *E. coli* 0157 at Kansas State started with cow-calf herds living on pasture and extensive range. Most of the previous on-farm work on *E. coli* O157 in cattle had been with more intensively managed dairy or feedlot cattle and we wanted to see if *E. coli* O157 was a problem in range cattle. The ranches were quite a change from my experiences on Ontario farms with small to mid-size dairy herds in peri-urban areas. The ranches in the beautiful tall grass prairie area of Flint Hills, Kansas are large, with extensive pasture and rangeland. Collecting samples required a fair bit of time roaming pastures on ATVs, and in some cases horseback, looking for the herd.

Dave Renter also collected samples from wildlife and the water on cow-calf ranches and he compared specific strains of *E. coli* O157 from different sources. He demonstrated that cattle and wildlife on the same pasture shed identical strains of bacteria which were also found in nearby free-flowing water sources. Clearly, the story of *E. coli* O157 in agricultural environments was much more complex than simply infected cattle. Indeed, later work in this area by the Kansas State team and other research groups has identified *E. coli* O157 in a variety of species, including flies and birds.

We were finding out if feedlot management affected fecal shedding of *E. coli* O157. We also wanted to know if *E. coli* O157 had economic consequences, even though the cattle didn't become sick. We started this study not long after the devastating outbreak of foot-and-mouth disease in the United Kingdom. With everyone's heightened awareness of biosecurity, we were worried that feedlot owners wouldn't want us to visit their farms and take samples. In the end, however, the owners of 73 feedlots in four states agreed to participate which speaks volumes to their confidence in us and their tremendous concern about *E. coli* O157.

The study took place during the summer months when the highest levels of *E. coli* O157 are found in cattle, and also when summer students can be hired to help. Although the techniques for diagnosing *E. coli* O157 were much better, they were still time consuming. Over that summer, we tested more than 14,000 samples of cattle feces, cattle feed and water from cattle drinking sources. To put this number into perspective, in the largest previous study of *E. coli* O157 in cattle, two laboratories examined about 12,000 samples over a one year period. Needless to say, the summer was busy, with six students in the field collecting samples and administering questionnaires and more than a dozen additional students working in the lab. The great organizational skills of Xiaorong Shi, our head laboratory technician, kept the whole process running smoothly. Once however, a courier lost a box of fecal samples and when it turned up several days later, the stench was horrible. I felt sorry for the driver who finally delivered the 'priceless' package! On another occasion, the enthusiastic students in the field collected more than 800 samples in a single day. The students in the laboratory worked well into the night to process them and even though I brought in pizza at midnight (after they finished with the fecal samples, of course!), it was days before any of them would look me in the eye.

Our study identified *E. coli* O157 in almost all of the feedlots, with about 10 per cent of the cattle shedding the bacteria. We also found *E. coli* O157 in their drinking water and feed. However, management practices didn't seem to make a difference in fecal shedding and no direct economic consequences were identified.

This work at Kansas State on *E. coli* O157 reinforced in me why I love epidemiology – how diseases occur, are transmitted and con-trolled – all of which can be applied to a wide range of problems and threats. In the end, we didn't find a solution to the *E. coli* problem, but we learned some valuable lessons. A common theme in *E. coli* O157 control had been to develop "on-farm" preventive strategies. Finding *E. coli* O157 in deer, other animal species and water was important because bacteria don't stop at farm borders. We would need to think more broadly about controlling *E. coli* O157 on multiple farms sharing a common ecosystem. While cattle certainly are the major reservoir, the epidemiology of these bacteria is much more complex than a single reservoir species. And improved management practices on individual farms don't provide the "magic bullet" to eliminate the problem. Recently some new developments have been encouraging - a vaccine for cattle is available in Canada and probiotics ("beneficial bacteria") appear effective in reducing fecal shedding. Clearly, the solution to the problems of *E. coli* O157 will take on-going research and the collabo-ration of the cattle industry. I'm gratified that our research group found a few more pieces of the puzzle.

The Night Roost

David L. Pearl

"Did I just pass County Road 5?", I thought as I was glancing back and forth between the road and a napkin where I had written down directions. I was visiting Wooster, Ohio for a few days to meet with research partners on a project about the role of European starlings in the transmission of *E. coli* O157:H7. It had been almost two years since we put the grant proposal together, and the fieldwork was finally underway. Jeff LeJeune, a microbiologist, was leading the research group, and for the next few weeks Jeff Homan and his group from the United States Department of Agriculture (USDA) were in town to conduct radio-tracking studies on the starlings. I was the epidemiologist in the group and was heading down to see whether our planned study was going to stay on track. It was fine to help set things up from behind a computer, but once the fieldwork begins and reality sets in, it's best to be around when the improvisation starts. Whether you're a field biologist or an epidemiologist, re-planning seems to be an occupational hazard. The animals or people never seem to behave the way you planned or provide the data you actually wanted or expected. On the other hand, if you're in the right frame of mind this can sometimes lead to unexpected opportunities.

"I made it!", I said out loud to myself when I realized I hadn't missed my turn. Jeff LeJeune said I had to see this night roost of starlings, but I had to get there before dusk or the show would be over. He said the directions were easy, but I was late setting out from the hotel where I was staying, and turns seem to come up a lot faster when you don't know where you're going. I almost ended up spending the evening in Akron, Ohio when I missed the first exit off the highway.

Jeff LeJeune and I met during my first job interview for a faculty position. He had picked me up from the airport to take me to the Wooster campus of The Ohio State University. Jeff was a member of the search committee and liked that my PhD research dealt with *E. coli* O157:H7, his pathogen of choice, and was excited by my work involving molecular and spatial epidemiology. I wasn't offered that job, but I guess I made a good impression. A few months later, after I started working as an assistant professor at the University of Guelph, he called to see if I would be willing to collaborate on his starling project.

I was finally heading slowly down a gravel covered country road when I saw the woodlot that the birds used as a night roost. I could see a couple of dairy farms further down the road, but the homes near the woodlot didn't appear to be part of working farms. I kept looking for a good place to park when I saw Jeff Homan's USDA car in the middle of a field about 500 meters from the woodlot, and decided to join him. The car looked like something out of a science fiction movie with a large antenna sticking out of the roof. I came up to speak with Jeff on the passenger side while another USDA biologist was focused on the radio receiver.

I met Jeff Homan for the first time earlier that day while he, Jeff LeJeune and a group of other USDA biologists were trapping starlings in mist nets at a dairy barn. Although we were both co-investigators on the same grant, Jeff Homan and I had never met in person. It was nice to finally put a face to his name. They had set up the mist nets, made of very thin black nylon, across a barn door. These nets were pretty tough to notice for an animal flying quickly through a tight space. Jeff Homan and I walked from one end of the barn making noise and flushing the birds to the other end, and hopefully forcing them into our nets. The day's catch was fairly good. Ten birds were fitted with radio-transmitters and released. The transmitters were fitted with little elastic harnesses for the birds' legs so that the transmitter would be held along their lower backs. It looked a bit like a jock strap in reverse if you ignored the wire hanging from the transmitter. The transmitters weighed only a few grams and were light enough to let the birds fly comfortably. Other birds were given colour tags on their legs so their farm of origin could be recorded if they were seen again in the study area, and some birds were sent back to the lab for microbiological testing. It had been almost 10 years since I had used a mist net. That project involved catching bats when I was a Master's student in Biology; long before I decided to become a veterinarian and an epidemiologist. Although I saw how feathers could add a bit of a challenge to getting these starlings out of a net, at least they didn't have teeth.

While handling these little birds, weighing about 60 grams and measuring 19 cm in length, beautiful flashes of iridescent green could be seen along their black feathers. Their short tails and triangular wings gave them a star-like appearance when they were in flight. Starlings number in the millions across North America and their common appearance in urban and rural settings makes it easy to forget that they are not native to this part of the world. Supposedly, they were first introduced to North America in the 19th Century as part of an ambitious, but misguided attempt to populate New York's Central Park with all the species identified in Shakespeare's writings. Their aggressive nature and adaptability gave them a distinct advantage over many native species of birds, and their habit of congregating in large numbers has made them a nuisance to many farmers. They are known to consume large amounts of feed and cost the US livestock industry almost $800 million per year. On this day, our interest in these birds had nothing to do with their appetites, but their potential to spread foodborne pathogens, especially *E. coli* O157:H7, among farms.

At first glance the role of starlings or wildlife may not seem obvious to the epidemiology of *E. coli* O157:H7. The major reservoirs for this zoonotic pathogen are ruminant species, and particularly cattle in North America. After the Jack in the Box burger scandal of the 1990's, the fear of contaminated ground beef captured media attention. The concept that someone in your family could go through bloody diarrhea, kidney failure, stroke and even death from undercooked beef is disturbing. While the beef industry worked to improve the safety of its product, and organizations like Pulse-Net helped use DNA finger-printing to help identify and recall contaminated products, the nature of *E. coli* O157:H7 as more than a "hamburger disease" was growing. The movement of the pathogen would be especially highlighted by the North American leafy lettuce outbreaks centered in "America's Salad Bowl" in California's Salinas Valley. While cattle foraged in the foothills above these fields, questions began to emerge about the route *E. coli* O157:H7 took before contaminating the lettuce. Everything from wild birds, feral pigs, and migrant workers were considered as being potential carriers during the investigation. Now, a bacterium that lived happily inside cattle was involved in ecological issues concerning water safety and wildlife, and social issues including the working conditions of migrant agricultural workers; it was becoming increasingly obvious that a more complete understanding of how this pathogen moved among farms and through the environment was necessary.

After working that afternoon, capturing and tagging starlings the big question arose about how we were going to pick other farms for radio-tracking. Before the visit, we had decided to randomly sample from a group of farms in Wayne County that agreed to participate in the study. "Random" was every reviewer's favourite word when you submit a manuscript for publication; random meant that there was little chance of the researchers introducing bias into the study. I can still remember sitting at my desk and using a random number generator to select the farms where we would band birds. However, once in the field it became obvious that it wasn't practical to have the farms too far apart since even with a car and a few permanent data loggers picking up radio signals at each farm, there would be little chance of picking up birds outside their home farm. Still, the researchers from the USDA were worried we wouldn't want to have farms too close together. In the end, we settled on a group of farms that were within realistic driving range of each other, offered access to birds, and were varying distances apart. It would be over a year before I saw the results of this radio-tracking study and our non-random choice gave us surprisingly strong insights on how the birds moved. They spent most of their day at this point in the summer, after the nesting season, centered around their home farms making only short trips to neighbouring farms, and while they quickly left their night roosts and made their way to their home farms each morning, they returned to their night roosts at a more leisurely pace.

Standing near that night roost as dusk approached, I was in absolute wonder of these birds. Thousands of starlings were circling this small woodlot. The birds looked more like a swarm of locust as they moved back and forth above the trees. It was impossible to imagine how they didn't crash into each other with each sudden wave of motion. Overhead, there were some phone lines that began to sway from the weight of birds waiting on these wires. While I watched in amazement, Jeff Homan left his car to join me. His colleague continued to sit in the car with a data logger and receiver hoping that some of the birds we radio-tagged that day would appear as little electronic beeps and be identified through the individual radio frequencies from their radio-transmitters.

As we sat watching, Jeff and I chatted about why these birds congregated in such huge numbers at night. While I wondered about issues concerning safety in numbers from night-time predators, he mentioned theories about information sharing. As the waves of birds above the trees continued, I asked him if he had ever gone into one of these wooded lots while the birds were "swarming". Surprisingly, despite years working with starlings and other nuisance birds affecting agriculture and airport safety, he had never gotten that close. I suspect any sensible person would want to avoid being covered in droppings, but on this night curiosity and a willing accomplice would make us take a closer look.

As we approached the woodlot, he warned me about a patch of poison ivy I was about to walk through; I was still too busy looking up in the trees, and taking short videos with my digital camera to notice anything on the ground. As we entered the woodlot, the leaves on the trees made it darker than expected and it was almost impossible to make out any birds at all except for the incredible amount of noise. Oddly, as we walked through this wooded area, the mixture of vocalizations from the thousands of birds sounded much more like a waterfall than anything avian. I was also surprised not to find many droppings at ground level considering the density of birds that must have been visiting this site each night.

As we headed out of the woodlot, night had fallen and we could make out the shadow of a little brown bat as it foraged for insects across the field. I remembered from my biologist days watching as female little brown bats and their young returned before sunrise from a night of foraging. Hundreds of these bats could suddenly be found circling around a small entrance in a rooftop, and a few minutes later they would all be inside and there would be no trace of them. Like the starlings, it was amazing how an enormous group of animals could converge daily in one place at dusk or dawn and be completely overlooked by the people living nearby.

The exciting thing about getting your head out of the journals and from behind a computer for even a few minutes is how many ideas you get from just watching the real world. As I drove back to the hotel, I was excited thinking about the data that would eventually come from our study farms where fecal samples from cattle and starlings were being collected for microbiological analyses. Would the farms with more starlings have more *E. coli* O157:H7? Could the birds be maintaining the infection among herds by contaminating cattle feed? Would the DNA fingerprints from these bacteria be the same in the cattle and the starlings that use the same farm? Could the birds pass this bacterium among each other at a night roost? It would take two more summers before all the field data were collected and a third year of laboratory work before we could begin the statistical analyses. It would be clear that despite putting together radio-tracking studies, laboratory studies, the collection of samples from cows and starlings on over 150 farms, and the completion of long questionnaires concerning farm management practices, we were only going to touch the surface of our growing mountain of questions. While our questions seemed infinite, it would also become painfully obvious that budgets and the agendas of funding agencies would have as much to do with our next steps as researchers as any discoveries we would make.

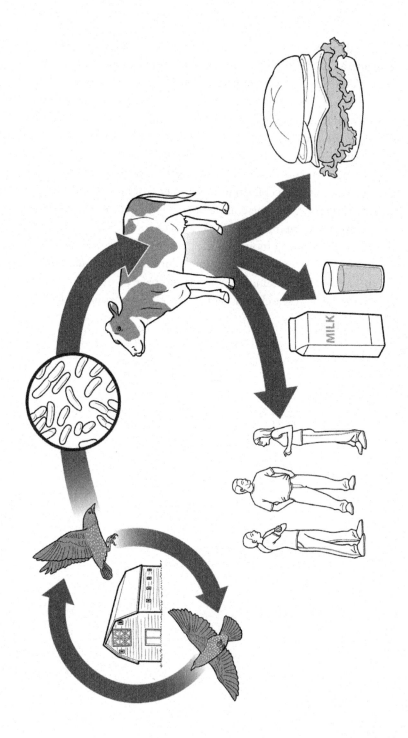

Figure 15 *E. coli* O157

What You Don't Know Can Kill You

Edward Breitschwerdt

I crashed to the floor on the few short steps from my kitchen table to the sink. Any doubts about the severity of my illness, were removed once and for all by the physical evidence of a gash on my shoulder and a lump on my noggin – not to mention the clear state of anxiety on my wife's face when she found me lying semi-conscious on the floor.

And all in all, I suppose I could consider myself fortunate because I had been felled by a bacterium I had studied for years. Not only that, the scientific investigation into my disease, Rocky Mountain spotted fever (RMSF) clearly demonstrated that RMSF can be transmitted by the lone star tick (*Amblyomma americanum*). Before my brush with danger, there were no reports in the scientific literature that the bacterium (*Rickettsia rickettsii*) that causes RMSF could be injected into a person by this tick species in North America. Let me hasten to add that while it was rewarding to successfully deliver such valuable evidence, in the future, I plan to avoid being the experimental animal!

Here is my story: I am a professor of veterinary internal medicine and for 25 years my laboratory has focused on infectious diseases that are transmitted by vectors, such as ticks, insects or other arthropods. Thus, compared to the average person, I know a lot about ticks and disease.

Along with my academic activities, my family and I operate an 80-acre farm, which makes for a pretty busy life for a 63-year-old. We keep about 20 beef cattle, and now that our two sons are through school and have jobs, my wife and I do the work on the farm. One weekend in late spring of 2010, I was doing this work, including pulling weeds from a hay field. A couple of days later, I noticed an itchy lump in my armpit and I saw an attached tick that was a little bigger than a flea.

There is a lot of folklore about the removal of ticks, such as using gas, kerosene or fingernail polish, all of which is probably wrong. If you can grasp the tick close to where it attaches, you can often pull it out gently, and that's what I did. I put the creature into a plastic bag, added alcohol and took it to our lab. This was important because if I became ill, we would need to test the tick.

I suspected it was a lone star tick because it had very long mouth parts that extended out from its body and a white dot on its back. My research technician looked at the tick with a microscope and confirmed its identity.

The lone star tick got its name because they were originally found in Texas, the Lone Star State, and other parts of the American South and because some people thought the white spot looked like the shape of Texas. Its range has now expanded to include northeast and central USA and southern Canada. The expansion of its territory is probably caused by the rapid expansion of the deer population in the USA and warmer annual temperatures. Twenty years ago we didn't have any deer on our farm and now it's common to come home and see half a dozen or as many as 15 to 20.

On a Sunday, about nine days after the tick bite, I was working with a friend on the farm. Usually by the end of one of these days, he is worn out, but on this day I was the one who was exhausted. You can picture it, a scorching North Carolina day, we'd stopped for a drink of water, and when I stood up I was dizzy. Let me tell you, when you are an old guy, "dizzy" scares you. Anyway, after a few minutes the dizziness passed, and I enjoyed my evening.

The next morning, after I arrived at the office, I was so cold, I turned on the heat (remember this is summer in North Carolina!). I am usually at the office pretty early, and by the time a research technician came in, she said 'this office is cooking'! I was a little miffed, noting I had been cold all morning. By that afternoon, I thought I was getting the flu.

Interestingly, even though I study tick-borne diseases and had had a tick in my armpit, I didn't initially consider a tick-borne disease! By the next morning, I had a fever, my muscles ached and I had a headache, which is a rare event for me. I asked a colleague to examine my blood and the results suggested a bacterial infection, not the flu.

I knew that lone star ticks can transmit other infectious diseases, but I didn't expect RMSF. I had a dead spot of skin in the centre of where the tick attached, which can be caused by RMSF and other rickettsial bacteria, but I assumed the dead spot was where I had squeezed the skin a couple times, causing trauma or a secondary bacterial infection in the moist environment of my armpit.

I asked a research colleague to check my blood and the tick for evidence of the DNA of *Rickettsia rickettsii*. And to our surprise, we found the specific DNA for these bacteria in both the tick and in my blood, thus proving a diagnosis of RMSF.

Now it was time to deal with my own disease. By then, it was the afternoon of the second day of illness, I had put in another full day of work and I was feeling really horrible. I looked up the correct dosage for doxycycline, an effective antibiotic for RMSF, arranged for a prescription and took my first dose.

I've only missed one day of work in over 25 years, which shows how sick I felt when I went home at 4:30, and crawled into bed. I gave my wife instructions about what to do if I had seizures and other symptoms of RMSF. When I woke about 10 pm, I was feeling much better as my fever and headache were both easing.

I hadn't eaten that day because I wasn't sure I could keep food down – people often vomit and feel nauseated with RMSF – but I was now feeling better and sat down in the kitchen. After a few mouthfuls of food, I became nauseated and got up to head for the sink, the short trip I never finished. Instead I ended up doing a faceplant on our floor and receiving the injuries I mentioned at the beginning.

After lying on the floor for several minutes, my wife got me back into bed. Enough food for one night! When I woke up the next morning, I no longer had headache or fever, but I had developed a rash. Medical texts state that if you have a rash, a fever and a history of tick attachment, you have RMSF. For me the rash was frightening because it emphasized just how severe an infection I had acquired from my "friend" the tick. The rash predominantly involved my arms and legs, and fortunately I had started the antibiotic soon enough. The rash worsened a little throughout the day, but started going away by afternoon.

Of course, there is ample irony in the circumstances of my story, but there is also truth to the adage that opportunity comes to the prepared mind. If that tick had latched onto someone else, it might not have been diagnosed as quickly. Without the diagnosis of RMSF, a doctor might have prescribed broad-spectrum antibiotics that are not effective against RMSF. And if the person waited for several days before seeking medical attention, even the right antibiotics might not have stopped the bacteria. I'm periodically called by lawyers asking me to be an expert witness in possible cases of misdiagnosis or mistreatment of RMSF; I always decline. If the patient does not report a history of tick attachment, the physician would not have RMSF on the list of possible diseases. The death rate from RMSF used to be 30 percent or higher before effective antibiotics were developed. And even now, 4-6% of people who contract RMSF die.

By the next weekend, I was pretty much back to normal (at least for a 63-year-old guy of German heritage), with no residual problems from my tick adventure or the bump on my head. There are numerous lessons to be learned from this story, particularly how critical it is to use tick repellents before heading out to the fields. Certainly, I was fortunate to have had one of the earliest interventions that anyone in North Carolina probably has ever received, thanks to my profession and my clinical and research colleagues. Of course, it is also fortunate that I understand ticks and it all gets back to the prepared mind: What you don't know can kill you; what you know can keep you alive.

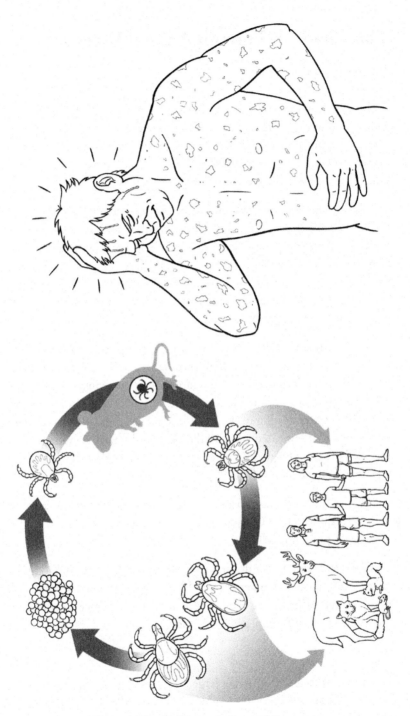

Figure 16 Rocky Mountain Spotted Fever

The Chaos of Foreign Animal Disease

Robert Curtis

In the 1960s, highly contagious Classical Swine Fever (CSF) was widespread in the United States. With no treatment available, officials in the USA initiated a widespread program to vaccinate millions of pigs to prevent the disease in the pigs that would be exposed to the virus. But the vaccines were not very efficient, so this did not work.

Classical Swine Fever was not nearly as pervasive in Canada as in the USA. The agency responsible for preventing an outbreak was Health of Animals (now called Canadian Food Inspection Agency). Canadian officials implemented a test and cull program to eradicate the disease from Canada. Because the results of lab tests for CSF look similar in pigs that have been vaccinated and pigs with the disease, vaccination wasn't allowed in Canada.

The Canadian officials required farmers to report any pig with suspicious symptoms such as high fever and diarrhea or if there were high numbers of pig deaths on the farm. After receiving a report, the government officials would come to the farm and take samples from all of the pigs for testing in the lab. If any of the pigs on the farm had test results that indicated CSF, all the pigs on the farm were euthanized. The government paid the farmers full market value for any pigs that were culled.

The test and cull monitoring program would have failed if any USA pigs were allowed into Canada, Thus, the border was closed to pig traffic. No pigs allowed!

Classical Swine Fever doesn't cause disease in people. In fact, pigs are the only species that get ill from CSF. But people play a major role in the spread of the disease and CSF can have huge effects on the emotional status of people as I will relate.

***

I planned to go into private large animal practice when I graduated from the Ontario Veterinary College (OVC) in May 1961. However, Dr. Blood asked me work at OVC for one year while he went to Melbourne, Australia to build a new veterinary college. I was pleased to have another year at OVC, to see farm cases of my own, and have professors nearby who would provide advice about my cases. I came to the farm service faculty for one year but stayed for 25!

I visited many pig farms two to three times a week. One of the farms in the herd health program had 200 pigs. The pigs were purchased from a sales barn when they were 6 to 8 weeks of old and placed in pens of 50 pigs each. The pens had cement feed troughs and automated water bowls. The owner did not live at the farm, so I was responsible to check the pigs regularly.

All went well until one morning when I found 10 pigs dead in one pen. They had seemed fine 24 hours earlier. Some pigs had diarrhea and were not eating or drinking. I loaded the dead pigs into the truck and took them to the OVC for post-mortem examinations. While the post-mortems were being done, I contacted the late Dr. C. K. Roe who was our swine expert at the time. Dr. Roe worked in several 'swine' states in the USA after graduation. Once he saw the severe diarrhea, dark blotches on the skin, and severe enteritis, he was sure we were dealing with Classical Swine Fever.

Pigs are infected with CSF in a number of ways, by ingesting contaminated food, inhaling droplets containing the virus, through natural breeding or artificial insemination with semen from an infected boar, by contamination of wounds or by the transfer of infection from a pregnant sow to her fetuses.

The virus can be spread from infected farms to neighbouring farms by birds, rodents, pets or other animals. Vehicles used to move live pigs, carcasses or manure may carry the virus as can people on their clothing, footwear or equipment.

The first outbreaks of CSF in Canada occurred in 1961 in Quebec; and in 1962, an outbreak was reported in Ontario. Dr. Roe was brought in to look at the affected pigs and he immediately diagnosed CSF. But how was CSF being spread and how could it be stopped?

The Quebec outbreak occurred in farm operations that used garbage from hotels and passenger ships to feed their pigs. Farmers trucked the garbage back to their farms where they were supposed to cook the garbage to a high temperature to kill all organisms. But, it was suspected that this was not being done. Officials traced the Ontario outbreak to pigs that had been brought in from Quebec to sales barns in Ontario.

Wherever the disease was found, the pigs were slaughtered and buried on the premises. Officials from Health of Animals handled the slaughter and they worked with the farmers to manage the overall situation.

Although my own work was primarily with dairy and beef cattle and sheep, I was fortunate at the time to have numerous discussions with Dr. Roe and with Dr. Ken Wells, the Canadian Veterinary General. I was fascinated by this outbreak and intrigued by how successful they were at eradicating the disease in a relatively short period of time.

By the end of 1963, Classical Swine Fever was eliminated from Canada. It took until 1978 in the USA after a 16-year program. Changes in the industry helped prevent any new outbreaks. Swine farmers are very careful about biosecurity on their properties. Rodent control is emphasized. As well, visitors are not allowed unless they put on special clothes, and on some farms, people shower before and after going into the barn. Very few pigs are sold through sales barns; farmers buy directly from other farmers. Furthermore, where garbage feeding is permitted, it requires a special license and federal government officials conduct periodic inspections to make sure the garbage is cooked to the correct temperature. In addition if pigs are taken to an event such as the Royal Winter Fair, they don't return to the farm, but are sent for slaughter.

The source of the infection in the 1960s outbreak was never identified, but it was theorized that the virus was brought into Canada on ships from other countries. To this day, outbreaks of CSF continue to occur throughout the world. The disease is endemic (consistently present) in some geographic areas of South America, Europe and Asia. Although CSF was eradicated from Britain in 1966, limited outbreaks occurred in 1971, 1986 and 1987 as a result of infected meat being fed in swill. This illustrates how widespread CSF continues to be and reinforces the significance of Canada's accomplishment in eradicating CSF more than 50 years ago.

In retrospect, as I recall the devastation that resulted from a CSF outbreak, I wonder about the emotional implications for farmers and their families. I have been unable to track down the exact number of farms or pigs involved, but Dr. Wells estimated that 1,000 pigs were destroyed at a cost of more than $1 million. No one talked about what was happening to the families, but the aftermath may have been similar to what happened in 2001 in England with Foot-and-Mouth disease (FMD; a highly contagious viral disease in cloven-hoofed animals).

To control FMD in England, millions of cattle and sheep were killed. It is sad to think about many of the farmers whose farms and the animal bloodlines that had been in their families for decades were lost. These farmers felt they had let down their ancestors and their families in addition to their heavy economic loss. Only belatedly did health officials recognize the mental health issues for the farmers when anecdotal evidence surfaced about farmers and others in rural communities suffering stress and depression - and attempting and carrying out suicides.

Chilling experiences affected the children as well such as witnessing the burning piles of dead animals close to their home and being sent away when their farm was quarantined. Many farmers were subjected to unprecedented emotional stress from the loss of their livelihoods as well as coping with the worry of the future of an industry in such a crisis.

In England, the long-term effects of the Foot-and-Mouth disease crisis will continue to have considerable consequences on the social health of farming communities and those living in the countryside for many years to come. It's a pretty sobering observation. We must consider the psychological, not just the physiological, effects of these outbreaks of foreign animal diseases. Classical Swine Fever and Foot-and-Mouth disease are not zoonotic, per se, but it seems clear to me that such diseases can pose a different kind of threat to humans.

Dead Chickens Keep Valuable Secrets

Dominique Charron

Maeh ('mother' in the Khmer language) was upset. Several more of our chickens had died that day, and the man who comes sometimes to the phum (Cambodian village) by moto (motorcycle) would not take the rest of the chickens to market because he was worried they would make his other chickens sick. Afterwards, Papa asked me to take our chickens and ducks to a secret place in the forest. We had six chickens left and 12 chicks that like to run free. We also had lovely fat white ducks, my favourite. I would take them to the rice fields and watch them search for bugs and weeds in the water.

A while ago in my phum, many chickens became sick and died. Papa and other men killed the sick ones for us to eat so they would not be wasted. Later, my friend Boupha, who was nine years old like me, became very sick. I didn't see her before her father took her to a doctor in the city. It makes me sad to think about her because she died there in hospital. The whole phum was sad. Maeh told me the chickens had made her sick.

"Kolab," she said, "you must tell me if the chickens are sick and I will look after them myself."

Later, Maeh was angry with Papa for having me take the birds to the forest because she worried they might be sick and a danger for me. But yesterday she couldn't take the chickens to the forest because she was selling baskets in Kranhung market with my older sister, Jorani. One day I hope to go to school in Kranhung, but it is far from my phum.

Nothing is the same in the phum since the man from the Department of Agriculture visited us after Boupha died. He spoke about a new disease that kills chickens and ducks and said farmers must inform the department when chickens or ducks become sick or die. If that happens, a person from the department will inspect them to find out the cause of the disease. So we ended up having more visitors than ever because chickens were always getting sick. After one of the inspector's visits, others returned and said that all our chickens and ducks had the disease and all had to be killed. Only Pu (uncle) Munni's prized fighting cocks were spared. They are beautiful, but mean and they do not produce eggs.

After that time, I stopped going to school because we no longer had the money to pay for my schooling, so I helped Maeh in the house and learned to weave baskets almost as nice as hers. I also learned how to choose the best chicks to buy so they will grow into strong healthy hens that lay many eggs. I helped choose our eight new hens when they were only little chicks and I helped raise them to be strong. Then one day I was quite sad because two of them died and only the day before they had been happy hens.

The dead chickens also meant a visit from the inspectors. I was worried the man from the Department of Agriculture would come to the phum to take away our new hens and ducks, so we hid them in the forest again. But if the chickens that died really had that terrible new disease, I would be so frightened because I did not want Jorani or my little brothers Phirun and Sovann to become sick like poor Boupha.

***

This morning, Maeh came with me to the forest to feed the chickens and ducks, kept in bamboo cages hanging in a tree. I love the forest but I do not go there by myself. Pu Munni shot a tiger there a long time ago and I am afraid of tigers. Maeh said the forest is our life. Without it, we have no fruits, wood, bamboo or medicines. We hunted birds and animals in the forest when we had no other food. We could find clean water there and good hiding places, too. There are areas in the forest with only grass and tall trees and this is where the chickens were hidden. We climb carefully because there is often a swarm of bees on the tree holding our caged birds. A few times in the year, we collected honey and beeswax. I love honey and it is a special treat. We kept some, but most of the honey is sold in the market. I was relieved because our chickens and ducks were surviving in the tree.

***

One afternoon I visited my friend Veata. I whispered to her about our chickens and ducks in the forest and how silly they looked high up in the tree. Then, under the family's sleeping platform, she showed me several small bamboo cages with her family's chickens! They are also hiding chickens. I thought this was better than hiding them in the forest, but Veata complained about the strong smell and having to sweep out under the platform every day. Soon, there was good news. One morning, Papa announced my birds could come home because no other chickens or ducks in the village had been sick, so it was now safe he thought. I hugged him and ran to get Veata to come with us to fetch my ducks and chickens.

***

There has been excitement in the phum. Even though there have been no sick or dead chickens, several people in a jeep came to visit us. Jorani, Veata and I ran from our houses to watch. They came all the way from the capital, Phnom Penh. It must be far, because their jeep was covered in dust and dirt. I noticed that there was a foreign lady with them. She must have been very important but we could not understand what she was saying. There was a friendly-looking man who seemed to be explaining what she was saying to Bawng (elder) Sokahn. They shook hands and then retrieved some items from the jeep. We wanted to see more closely but Maeh called us back to the house.

***

Later, Jorani and I lay awake, listening to Papa chat with Bawng Sokahn and other men while they sat and smoked in front of the house. They were talking about the people who visited earlier in the jeep. The men were from universities in Phnom Penh and Thailand. The foreign lady was from Canada. Jorani told me this is a cold country on the other side of the world, near America. Why would she come all the way to our village? Bawng Sokahn said they are experts who want to protect the phum from the new bird disease. Papa was suspicious and worried that the jeep people might make trouble for us. Bawng Sokahn agreed, but said the jeep people could help the village deal with the Department of Agriculture. He said the foreign lady and the friendly-looking man were both doctors for animals.

Doctors for animals! How wonderful! I was so excited I almost jumped off our sleeping platform to run outside and ask questions, but Jorani hushed me. I whispered to Jorani that one day I will become a doctor for animals.

"Kolab," Jorani said, "your head is full of impossible dreams."

I ignored her while thinking about helping ducks and chickens as a doctor for animals. And perhaps even the poor old dog with the sore foot, but not Pu Munni's fighting cocks. They were too mean.

***

I thought we must have been the luckiest phum in Cambodia because the jeep people had returned! Maeh let me listen to them and she sent me with a tray of hot tea for the visitors. I met Dr. Kiri Bonarith. He is from the capital, and he told me that a doctor for animals is called a veterinarian. He was very nice to me and explained things in simple words I could understand. With his jeep friends, he wanted to help farmers and villagers learn ways to protect our animals, and make them grow bigger and healthier. He said we would also protect ourselves from some diseases, such as the bad new bird disease called influenza. The foreign lady veterinarian from Canada was Dr. Kathy. She worked with other Canadian veterinarians who belong to Veterinarians Without Borders Canada, and they travel around the world to help people and animals in poor countries like Cambodia. The other person with them was from Thailand, but he spoke a little Khmer. He was called Professor Djum and works in a university in Ubon.

Dr. Bonarith and Dr. Kathy toured the village and I went with them. Bawng Sokahn explained there were 30 families in our phum with a variety of animals. Each family had chickens and many had ducks. He did not know how many birds we had then, but it was fewer than before because of the influenza outbreak months ago. Some families also had one or two pigs, but there were only 10 pigs in the whole village. There were several water buffalo used to work the fields and sometimes to transport heavy things on a cart.

Dr. Bonarith and Dr. Kathy then visited the animals. This was more interesting and I tried to make myself useful. First, we visited Veata's house. Veata's father listened for some time to Dr. Bonarith, then disappeared into the house and returned with a dead chicken. I was curious to see what the veterinarians could do for a dead chicken. Dr. Bonarith explained veterinarians could not help a dead chicken, but they could sometimes help keep others alive.

"Dead chickens keep valuable secrets," he said. Veata's father was suspicious at first, but then changed his mind and gave them the dead

199

chicken. Dr. Bonarith said Dr. Kathy would examine the chicken to determine what may have killed it. This sounded like magic to me.

Dr. Kathy put on rubber gloves and a paper mask, and spread a plastic sheet on the ground. I was not allowed to come close. She produced a pouch with shiny tools, a bottle of disinfectant, some small jars and tubes. She turned the chicken this way and that, looking into its eyes, ears and beak. She felt it all over and looked at its skin. She then opened the chicken with a knife and used shiny scissors to cut bits of chicken, placing them in the various tubes and jars. I was eager to know what she had seen in the poor dead chicken.

Dr. Bonarith called me over, without letting me come too close, and showed me where the chicken was thinner than it should be, and that it had crusts in its beak and ears and around its eyes. A thin chicken like this one had been sick for a while. The crusty beak and ears meant it probably had a common disease of the throat and lungs, and not influenza. Although this disease was dangerous for other chickens, it was likely that many other village chickens had already survived this disease and were now protected from catching it. Why, I asked him, did Dr. Kathy take small pieces from inside the chicken? I was amazed to hear she would bring these back to the capital and look at them under a microscope. He told me that with a microscope we can see things so tiny that normally they are invisible. With the microscope we will find out what might have killed the chicken. It was magic, I knew it!

***

One day the man on the moto, Samnang, came to the phum to see if there were any chickens for market. Our jeep friends were interested in talking with him. I learned that he takes the chickens to a large live market east of our phum. Maeh told me that when we buy animals that are still alive, we know that they are healthy and fresh. We can also buy new animals there, such as baby chicks, pigs or buffalo. Samnang said that in the very big live markets, such as the one in the East, there were all kinds of animals for sale, such as snakes, lizards, leopards, screaming monkeys and a rainbow of song birds. There were many kinds of ducks, chickens and geese and pens with buffalo, pigs and cows. There were butchers who will kill and skin the animals, cut up the meat and wrap it in neat paper packages.

Professor Djum asked Samnang where else he goes on his way to the live market.

"Many phum, many *srae* (farms)," Samnang answered, and explained that chickens from the phum fetch the best price in the market. Rich people in the city miss the taste of the countryside, he says. So it was worth his trouble to ride around the countryside buying a few chickens from the villages, rather than buying chickens from big farms. Dr Bonarith translated for Dr. Kathy, who wanted to know more about the conditions for chickens in the market.

"When I arrive in the market," Samnang said, "I always try to go to the same area, because my best clients know to find me there. I often see other familiar faces selling this kind or that kind of bird, but also many new faces all the time." Usually all the chickens were mostly in one area, but sometimes there were sellers who have many kinds of birds, ducks, geese and wild birds, all mixed together. As he spoke, I thought of the bad bird disease, and so did Dr. Kathy, because she asked: "How many birds die, and how often, before you can sell them?"

Samnang thought for some time before answering, "my clients know that I have good healthy birds."

Dr. Kathy was still curious about diseases, and she asked about the health of other vendors' chickens. That was a different situation.

"There are always chickens dying," he said. Some vendors were not as careful as Samnang and they sometimes brought sick chickens to the market. The more careful vendors did not like this, and sometimes tried to prevent the bad vendors from selling at the market, but they were not always successful. Dr. Kathy asked about influenza and how it was in the market during the outbreak. Samnang shook his head and said many birds died in the market in a short time, and people stopped coming for a time, clients and vendors both. There have been new rules, but they had not changed the market very much, except that it was a bit cleaner and perhaps a bit less crowded because there were more spaces between the cages. The only real difference was that the butchering area was separate from the live animals. Dr. Bonarith then asked about the ducks and how they were kept. Although they were mostly in a separate area, there were ducks and chickens together all over the market. Later, Dr. Bonarith told me that ducks can have influenza without becoming sick but the chickens can catch influenza from the ducks and get very sick. More magic and mystery! Now the ducks had secrets.

***

Dr. Kathy visited each house in the village to learn about all our animals. She opened more dead chickens and some ducks, and one day I was amazed to find her behind Veata's house with her arm buried up to her shoulder in a buffalo's behind. I know that sometimes it is necessary to help a calf come into the world, but Veata's old buffalo has not had a calf for a long time. Dr. Bonarith explained Dr. Kathy was trying to find out why Veata's buffalo had not had a calf. I did not think she would find the answer in the buffalo's behind, but actually the buffalo had a problem inside because of her last calf. Dr. Kathy could give her an injection that may help the buffalo become pregnant.

Another day, Dr. Kathy visited Bawng Lay's house to see his pigs. He has a big house and the most pigs of anybody in the phum: two sows and a big smelly boar. We visited the pigs outside in their pens. Dr. Kathy wanted to know how many piglets had been born and how many survived. Bawng Lay is an old man, but he knows a lot about pigs. He said his sows often have two litters, and he was able to sell many piglets to others in the village or to market. But one sow would often lie on top of her piglets or bite them and they do not all survive. The doctors inspected the pens where the sows were kept when they had their piglets. Dr. Bonarith explained the pens can be changed to let the piglets escape from under the sow, but still be close to her to keep warm or to have access to her milk. He drew in the dirt to show Bawng Lay how to build this kind of pen.

***

Then one afternoon, it seemed everyone in the phum gathered under the Banyan tree, or at least all the grownups. Bawng Sokhan had called a meeting of the village. Dr. Bonarith, Dr. Kathy and Professor Djum were there, too. Dr. Bonarith explained our phum was part of a study with many other villages in several Asian countries, including China, Thailand, Indonesia and Vietnam. In all these other phums, people like us were also trying to protect their animals and families from influenza. Some other professors were studying where the wild birds go and how they might have been catching influenza from chickens or ducks and spreading it to other places. Dr. Bonarith had other professor friends trying to understand how the Department of Agriculture makes new rules and whether these new rules make any difference. These friends were called the Asian Partnership for Emerging Infectious Diseases.

Dr. Bonarith asked us to think about how we might protect our chickens and ourselves from the disease. He suggested keeping other animals instead of chickens and ducks, keeping our chickens and ducks separate, or keeping the birds out of the house. The grownups talked in groups. Pu Munni talked about fighting cocks with other men. He kept them in cages but he sometimes took them to other phums where they battled other fighting cocks. He wanted to give them a medicine to protect them from catching influenza. Another group of men discussed ways of keeping ducks and whether to take them to the rice paddies. Maeh talked with a group of women about chickens. In my phum, women and girls looked after the chickens. Dr Bonarith talked about how chickens are killed and how to be careful not to catch influenza. Maeh and Veata's mother argued about letting chickens in the house or not. Maeh did not like the chickens coming in because they were messy. She always shooed them out with a hiss and clucking sound. Veata's mother said her chickens are fatter than Maeh's because she lets them roost in the house at night. She said it was safer and she loses fewer eggs this way.

Later, Dr. Kathy was sitting in a folding chair near her tent, writing in a notebook.

"Sua s'dei!" she called to me and I answered "hello," giggling that we have traded languages.

"Daa neh chmuah ey?" she asked ("What is your name?").

"Kolab," I replied, and I really wanted to tell her about my dream to one day become a veterinarian, but I did not know any more English words so I said it in Khmer.

"Ot toh," she says with a sad face, "minh yol dtey." (Sorry, don't understand.)

She waved to Dr. Bonarith for help. I was suddenly shy and did not know what to say. He sat on the ground so he was even smaller than me. Finally, I told him about my dream. He smiled while telling Dr. Kathy what I had said. She took my hand and said some words. Dr. Bonarith explained that I will need to study hard at school and have very good grades to go to college. He told me there is no college for veterinary studies in my country, so I will also need to learn another language like English to be able to study elsewhere. But, he said, my country needed many veterinarians.

***

Then one day I was quite sad. Dr. Kathy, Dr. Bonarith and the professor left in their jeep to return to the capital. Dr. Kathy gave me a new notebook with white paper and some pencils for my lessons. I will save them for when I return to school. I met Veata and I told her why I was moping, about my new notebook and pencils, and I even told her about my dream to become a veterinarian one day. She said she would not be surprised if my dream came true. She also said the jeep people have not finished and will return after the rains.

***

By then, we were raising many more chickens, since many of our chicks had grown. They were good egg-producers and had been mostly healthy. In addition, there have been no outbreaks or dying chickens since before the jeep people arrived. Since we were doing better, Maeh said I could return to school at the beginning of next school-year, after the rainy season.

***

Just the other day Maeh went to Kranhung to fetch my new school uniform. It was the best day ever! I was accepted into the school for girls because I am clever enough, and in a few weeks I would go away to school in Kranhung. Shortly after, Dr. Kathy, Dr. Bonarith and the Thai professor returned. I cannot wait to tell Dr. Kathy and Dr. Bonarith that I will soon be going to school!

***

I listened carefully to Dr. Kathy to try to learn some English, but it was hard. She was using words like "laryngotracheitis" and "vaccination" and "pasteurella," which sound more Francais than English to me. The doctors sat under the Banyan tree with Maeh and many women from the phum and explained what they found with the microscope. The chickens will not hold their secrets for much longer. Dr. Kathy said many of our chickens and ducks were too thin because they were full of worms and were not eating enough. She also said the chickens' bones were thin and that we should feed them more eggshell, seashells and snails. She then talked about what may have killed some of the chickens. We were relieved it was not influenza, but the common disease that caused crusty eyes, beaks and ears. However, two other more serious diseases were also found. Dr. Bonarith was surprised more chickens had not died from a disease called Newcastle. It is very contagious and a big problem in many phums. Dr. Kathy also found a disease called fowl cholera in some of the young duck-lings and chickens that had died. Maeh looked worried about this, but Dr. Bonarith said quickly that this is not the same cholera that killed people. Dr. Kathy said there were medicines that can prevent these diseases but they often need to be given by injection. They are expensive and difficult to get to the phum. But there was a new medication that can be given in the beak or nose of both ducks and chickens to prevent the cholera. There were also inexpensive medi-cines to reduce the worms and help the chickens grow better.

Later, the grownups gathered again to listen to the veterinarians and the professor. Professor Djum said he was surprised how many more chickens there were in the village. He explained that, even though there are still problems with influenza all around, and not just in Cambodia, farmers in phums recover quicker than those at bigger farms. He said that this was because our families do many different things and raise many different animals and crops, so there is some-thing else to eat or sell for money.

Dr. Bonarith talked about ideas to help prevent influenza and other diseases in the phum. Maeh and Veata's mother suggested we bring our chickens and ducks to one place near the road, rather than wait for the men on motos to visit each house to get chickens. That way, if there is a sick chicken in the village, it does not visit all the houses. Dr. Bonarith agreed and they talked about where to gather the birds and what to do in the rainy season when there might be too much water by the road. The doctor asked if it might be possible to keep the ducks near the phum and away from the paddies with the wild ducks and geese, which would help limit how far influenza can spread.

Professor Djum met with Bawgn Sokhan, Maeh, Papa and others and talked about growing more animals and crops and finding clever ways to sell them for more money. He said it was possible to borrow a small amount of money from special banks for poor farmers, sometimes only for women, to buy things such as metal wire for fences or cages, seeds or animal feed. The bank is run by other poor farmers. The money needs to be paid back once the animals mature and are sold or the crops are harvested. This was a new scheme in our area. The professor also talked about how a chicken becomes more valuable the farther it goes from the village. Men like Samnang sell the chicken for more than what they pay us for it, as do the men who buy the chicken from Samnang at the market.

***

I am late for my class! How can I expect the students to be on time if their professor can't manage it herself? Perhaps they will be held up in their surgery labs and I will have a few minutes to get it together. This is a new bunch, and I need to grab their attention quickly with this ecohealth lecture.

"Welcome, class. My name is Kathy Williams-Tanner and today we're going to take a little trip to Cambodia. Along the way, I hope you'll see the connections not only between animal health and human health, but also between ecosystems, social and economic activities, and health."

It's been almost eight years since I was in little Kolab's village. When I last saw her, she was busting to tell me about her new school and her uniform. I guess she'd be finishing high school or even at college by now, if she was fortunate enough. I wonder if she's still learning English.

Dr. Bonarith stayed in touch with the villagers for a while. He went back the following year with Djum, and things were coming along very nicely. They had an entrepreneurial mindset in that phum! He wrote that they are still farming traditionally, but had made several minor improvements to their poultry housing. They had also developed a new marketing scheme for their high quality village chickens and ducks, targeting Siem Reap and even Phnom Penh. The micro-credit loans made a big difference for the women in the village, who were able to buy de-wormers and vaccines, and then generate higher egg and meat production. Since they keep most of the proceeds from things like chickens and baskets, this added income has a big effect on the quality of life for their kids. They had bought bed-nets to prevent malaria, and more kids were in school. They had even arranged for a mobile health clinic to come regularly to their remote area. Cambodia did have another outbreak of avian flu since then, so I hope they were not too badly affected.

***

Today, I was a bit nervous. I had my first English tour group at Angkor Wat. I had learned all about the temple and its history and was ready to answer questions. I have practised my English and my teacher said I was a promising student. I was saving all my money to help Maeh and Papa pay for school for my brothers Phirun and Sovann, and to pay for college. I would like to attend the Prek Leap National School of Agriculture. If I do really well, then I may be able to earn a scholarship to study veterinary medicine. I have looked online and there are so many schools to choose from. But I think that if I have a choice I would like to go to Canada.

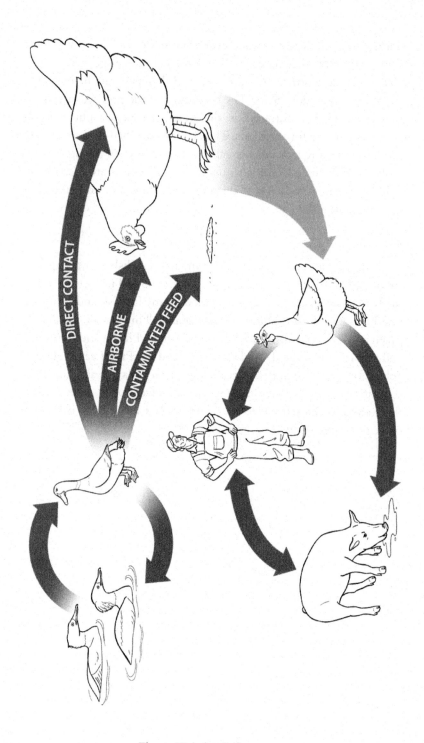

DIRECT CONTACT

AIRBORNE

CONTAMINATED FEED

Figure 17 Avian Influenza

The Koi and the Cage: A Veterinarian's Personal Medical Odyssey

Greg Lewbart

"I'm afraid amputation is the only option for you now," said the surgeon in a sombre tone. "If we don't, you could lose your arm. I'm sorry."

"Doc, so you're telling me that after nearly a month of chasing after this infection, thousands of dollars in medical bills, and countless hours of lost time, I'm going to lose my hand anyway?"

"I don't know what else to tell you Rick," the defeated surgeon said. "I wish it wasn't so and that we could do more for you."

"You won't take my hand, at least what's left of it anyway," the stubborn veterinarian insisted. And with that comment Dr. Raymac walked out of the mid-sized regional hospital in search of a better option.

***

I had heard part of Dr. Rick Raymac's medical odyssey because we had talked over the years at various pet fish medical conferences. But I needed more details to write this story, so I asked him to send me a copy of his medical records. When the package arrived -- 18 pounds of clinical documents, legal briefs and a sealed plastic container holding a string of methylmethacrylate beads impregnated with antibiotics – I was astounded! He had warned me the package would be heavy, but this mass of ink-filled paper was physically and mentally intimidating. How could I possibly distill all this information into a coherent story that was accurate, educational and interesting? I decided to use a good old-fashioned recipe for New England maple syrup: Boil down the stacks of crude paper sap to produce, I hope, a pure and palatable story.

***

Rick, a veterinarian in a quaint New England town, practices small animal medicine and enjoys the outdoors, wildlife rehabilitation, and adventure travel. Perhaps his greatest interests are his indoor and outdoor ponds of koi fish. One fall day, Rick was working in his well-manicured backyard, caring for his precious pet fish, and building a new outdoor cage for a family of orphan opossums he was nursing back to health. The young marsupials had been living in his daughter's doll house and they needed more space. As well, they needed a safe and secure outdoor habitat where they could acclimate to the environment and to a normal day and night light cycle in preparation for their release in a few weeks. Rick was even including an upright tree trunk complete with a shelter hole and several branches for just hanging out.

Suddenly, he let out a short, audible curse as a small, needle-sharp piece of the galvanized metal caging material slipped free of his grasp and sliced deep into his left ring finger. He yelped not so much because of the pain, and it did hurt, but because he knew he should have been wearing the work gloves that were just a few feet away in his garage. He quickly extracted the metal, which was just 1-2 millimeters in width and about a centimeter long. He tossed the tiny dagger aside, and after putting away some tools and supplies, went into the house to wash up. He didn't realize a small piece of galvanized metal coating remained in his finger, lodged securely inside the sheath of the main tendon.

Two days later after the opossum cage incident, the end of Rick's finger "blew up like a balloon...very red, very painful." Recognizing infection had set in, he started himself on the antibiotic cefalexin.

The next day, with his finger still swollen and painful, Rick visited a hospital emergency room, where Dr. "N," a board-certified hand surgeon "with a big attitude," examined him. Although X-rays of Rick's hand revealed a metal fragment near the base of his infected finger, Dr. N insisted the infection couldn't be related to the cage injury and started treating Rick with antibiotics intravenously.

Dr. N operated on Rick's hand to clean out any unhealthy tissue and to irrigate it with sterile fluid. He also placed small methylmethacrylate beads that had been impregnated with antibiotics into Rick's hand. The slow release of antibiotics from the beads and the infusion of antibiotics into his vein started to control the infection and after six days of treatment, Rick was discharged from the hospital.

Two weeks later, Rick's entire hand, all the way up to his wrist, "blew up" again. It was at this point that Dr. N insisted the hand be amputated and Rick walked out.

***

The next day Rick had an appointment with an infectious disease specialist and another hand surgeon at a university medical center. The experts exclaimed, "Don't worry, it's just TB." Another "clean-up" surgery was performed, a different antibiotic (minocycline) was administered and within 24 hours Rick was much better. A diagnosis of a bacterial infection with *Mycobacterium fortuitum* was made once the results from biopsies and cultures taken during surgery were reviewed.

Unfortunately further complications developed – the infection spread to the bone. Over the next five months Rick made dozens of visits to doctors' offices and the hospital. He needed four more major surgeries, including two surgeries to take tendon tissues from one of his feet and graft them into his hand. These complicated, intricate procedures necessitated diligent care and monitoring of both his hand and donor foot.

***

Where had the infection come from? In retrospect, the source of the infection was most likely one of his newly acquired koi that had died shortly after he purchased it. Being a veterinarian, Rick did a necropsy (autopsy) on the three-inch fish and had seen internal lesions compatible with the disease mycobacteriosis of fish. He hadn't worn gloves during the necropsy, but afterwards he washed his hands thoroughly with antibacterial soap. Then he started work on the opossum cage and within minutes the metal jabbed into his finger.

A freshwater or marine fish infected with *Mycobacterium fortuitum* may exhibit chronic illness and in many cases will die. Affected fish frequently have skin ulcers, their eyes may bulge and they may lose weight. Sometimes a fish has no obvious signs before it dies. Infected fish with open sores spread the disease to other fish and may shed bacteria in their stool. Some veterinarians and others try to treat this disease with a variety of antibiotics, but the fish usually die.

The common name for this infection in people, and the term widely used by physicians, is *fish tank granuloma*. The disease nearly always includes a history of the patient being in contact with water, aquatic animals, or both. In almost all instances, there is a wound or opening in the skin. Unlike other species of mycobacterium such as those that cause tuberculosis or leprosy, there is no evidence that *Mycobacterium fortuitum*, or the closely related *Mycobacterium marinum*, can spread from person to person.

Many primary care physicians have little to no experience with the disease, and as such wouldn't consider mycobacteria as a possible cause of an infection such as the one in Rick's finger. Fish tank granuloma is rarely fatal to humans or other mammals because it almost never spreads beyond the local area.

***

Nearly ten years later, Rick continues to experience chronic but tolerable pain and reduced mobility and permanent flexion of his left ring finger. Certainly, things could have been worse, and acting as his own advocate, he was able to save his hand. The numerous surgical procedures, recuperation intervals, doctor visits and efforts at rehabilitation resulted in medical bills totalling more than half a million dollars and many months of lost work and income. As a result of all these challenges, Rick wants to educate others about the risks of mycobacteriosis and was eager for me to share his story and medical odyssey.

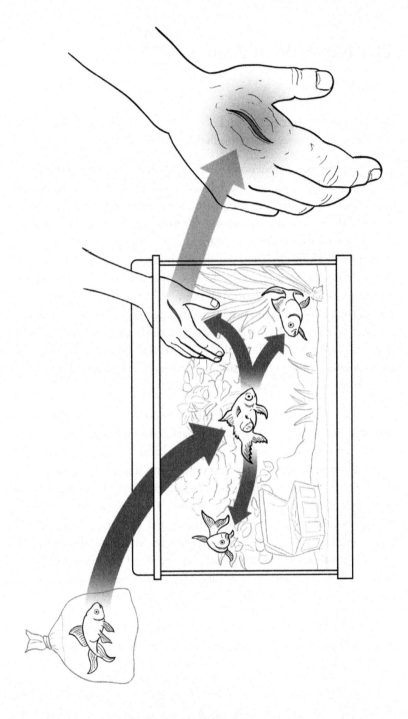

Figure 18 Mycobacteriosis

The Flip-Side of Zoonoses

J. Scott Weese

"We are not a population of dogs, cats, horses and cattle living amongst people... we are a population of animals."

A dolphin died of pneumonia. Tests showed that a bacterium called methicillin-resistant *Staphylococcus aureus* (MRSA) had infected its lungs. The veterinarians at the marine park called me because I have studied MRSA and other zoonoses for many years. After more tests on the bacterium, we discovered that it was a strain commonly found in people (CMRSA-2). The odds of this human strain occurring in marine mammals independently from people are exceedingly low, so it's almost certain this human strain of MRSA was passed from a person to the dolphin (or perhaps from a person to another marine mammal to this dolphin). MRSA is a type of bacteria that are resistant to the antibiotic methicillin and if left untreated, MRSA infections may develop into serious, life-threatening complications such as infection of the bloodstream, bones and or lungs.

As not much is known about MRSA in dolphins, we were concerned about the health of other animals and the people that come in contact with them. Thus, we immediately launched an investigation. Typically, these types of investigations in land mammals include screening of healthy animals (and people) to detect carrier animals that aren't sick but have the MRSA bacterium. The majority of individuals who encounter MRSA don't get sick, but may end up carrying MRSA, usually in their noses, for variable periods of time. We collect samples from people and most other species by swabbing their noses and sometimes the rectum. Investigating MRSA in marine mammals raises some interesting questions such as:

How do you collect nasal swabs from animals that don't have noses, such as dolphins and whales? You don't – you have to use their equivalent of a nose, the blowhole.

And even if the animal does have a nose, how do you convince a walrus, for example, that it should let you stick a swab up its nose? Patience and a quick hand!

214

Fortunately, veterinary personnel at the marine park were able to collect the samples - from dolphins, orcas, belugas, walruses, sea lions, seals and also the people who worked with these animals. In the first round of testing, MRSA was found in two of the six dolphins and one of the six walruses. This prompts more questions.

How do you isolate a marine mammal infected with MRSA from the other animals when they all live in the same circulating water? That's tough. With dogs and cats, it is fairly easy to restrict contact with other animals and people for a short period of time, even while keeping the pet in the household. Horses can be housed alone in a stall. Birds can be kept in a cage. What can we do when there's the potential for MRSA to spread via water? We weren't able to come up with a perfect solution, but instead implemented something that was practical and worked. The walruses were moved to a separate area of the park. As this wasn't possible for the dolphins, the infected and non-infected dolphins remained together. We realized this situation might mean ongoing infection with MRSA among dolphins, but we hoped that if we intervened when we detected a problem, the outbreak would run its course.

How do you keep people from becoming infected or from spreading infection to other people or animals? Park personnel were encouraged to follow basic infection control precautions, such as limiting contact with the animals and avoiding close contact with the noses and blowholes of the marine mammals. Particular attention was paid to personal hygiene, namely hand washing and the use of alcohol-based hand sanitizers. Our goals were to prevent infected people from infecting other animals, to prevent people from becoming infected from an animal, and to prevent people from contaminating their hands or clothing with MRSA and then passing it on to other individuals.

How do you get rid of MRSA in a marine mammal?

In people, therapy to eliminate MRSA from carriers, which is called decolonization, has involved putting an antibiotic cream in the person's nose, sometimes in combination with other treatments such as antiseptic baths or oral antibiotics. The effectiveness and necessity of this approach for MRSA carriers in the general human population is debatable and it's less commonly done now than in the past.

With marine mammals, can we really treat the nose or blowhole adequately? Won't the antibiotic cream simply wash off in the water? What's the risk of MRSA exposure to the person leaning over a blowhole and applying an antibiotic ointment? How do you bathe a dolphin? All of these concerns and issues made us question the idea of decolonization therapy in these marine mammals. Eventually we decided not to treat the dolphins because in many animal species the MRSA bacterium is not really adapted to live on the animal long term and the bacteria eventually disappear. We didn't know whether MRSA was transient in marine mammals, but given our limited options we went ahead based on this assumption and simply monitored the animals and took steps to reduce the potential for cross-infection as much as possible.

Over the next 20 months, we continued to test the animals at the marine park. Three months after the investigation began, the swabs from the walruses were all negative for MRSA and we never found MRSA in them again. The dolphins were a different story. In one dolphin, we identified MRSA for many months and we were concerned that she could be a persistent carrier. Fortunately, she eventually became negative after almost 18 months and we didn't find MRSA in her again.

The period of study and testing was longer than we had expected, but in the end we demonstrated that MRSA can be contained in a unique environment such as a marine park and that these animals are able to get rid of MRSA on their own. Too often, the tendency is to treat with antibiotics, despite a lack of evidence that this approach is needed in all situations and the fact that using antibiotics can lead to further antibiotic resistance. In this case of marine mammals, patience was the key, not antibiotics or anything fancy.

The origin of MRSA in this outbreak was never determined. A staff member who was a carrier could have been the original source because a small percentage of healthy people carry MRSA in their noses. In our investigation, we didn't identify MRSA in any personnel, but our testing was conducted well after MRSA had been established in the animals – and by that time, a person who had been carrying MRSA might have already cleared it. Another potential source of infection is the general public. During animal shows, this facility, like many similar marine parks, permitted people to have supervised contact with certain marine mammals, such as dolphins. Typically, a child from the audience gets to meet a dolphin during each show, which means that over time the dolphins come into contact with a large number of people. Thus, even though only a small percentage of the general population carries MRSA, there is a high likelihood that the animal was exposed to a MRSA carrier.

Once the last dolphin became MRSA negative, the facility was MRSA-free. However, the close interaction of the marine mammals with people will presumably continue to put them at risk of MRSA infection. The biggest thing that can be done to reduce the risk of human-to-animal transmission is hand hygiene, hand washing or using an alcohol-based hand sanitizer. This facility still allows public contact with the dolphins, but people have to use a hand sanitizer first. Next time you're at a marine park or other facility that allows animal contact, think about MRSA and look to see what steps are being taken to protect against it – for the sake of both the animals and the people!

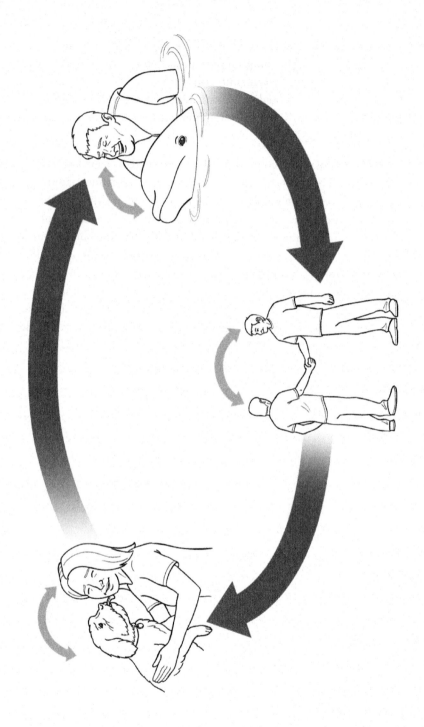

Figure 19 Methicillin-Resistant *Staphylococcus aureus* (MRSA)

Parasite's Log: The Ultimate Mission in Northern Canada

Andria Jones-Bitton

The day is cold and crisp in Northern Canada. The sun is shining making the tiny snow crystals sparkle as the wind carries them through the air. In the distance, the sound of snowmobiles signals a group of Inuit hunters on the move. Foods must be caught to feed and support their community through the long, encroaching winter. The seals, sunbathing on the edge of the water, decide today is not the day to meet with the hunters and they quickly roll into the cold water. Although calm and peaceful on the surface, below the surface the aquatic ecosystem bustles with its daily tasks. One of the most practical and necessary – a whale defecates, and in doing so, releases stools containing hundreds of thousands of tiny parasite eggs. And with that simple act begins the story of our protagonist, a parasite by the name of *Anisakis simplex*.

Parasite's Log

Day 1

Not the most glamorous entry into the world, I realize, but such is my fate.

I have been living as an egg in the intestine of this whale that has been my host. And now along with its stool, I'm pushed into the ocean to begin my Ultimate Mission - complete my life cycle, mature into an adult *Anisakis simplex* worm and reproduce. It's not so strange a life goal, is it? Surely many species take on similar missions?

The strength of my motivation for this quest is astounding. Numerous steps are ahead of me, each with its own hurdles and challenges, but I must remain true to my cause. The very life and survival of our species depends on it.

Day 16

Inside this thin-shelled egg, I continue to mature. The egg slowly moves in the cold water and is so small, you would need a microscope to see it. I must remain patient.

Day 27

Finally, progress to the next stage! I hatch from my egg as a second-stage larva. Now I can swim freely.

To advance into my next stage, I must be eaten by a crustacean – a shrimp, a krill – it doesn't really matter which type, but it is only inside them that I can mature to my next phase of life – the third stage larva. And that will bring me yet another step closer to my Ultimate Mission.

In these icy temperatures, I can survive for six to seven weeks – we *Anisakis simplex* are quite adapted to the cold. So here I wait, drifting in the cold water currents, waiting, hoping . . . to be eaten!

Day 41

Success! I have been eaten by my host of choice, a crustacean - a magnificent shrimp to be accurate. I can now develop into a third-stage larva. One step closer to my Ultimate Mission.

Once again, I will need to be eaten, this time by a fish. Of course, my host the shrimp must also be eaten - not a great end for such an accommodating host, but such is the way of the food chain. I'm not particular as to what kind of fish – a sculpin, polar cod, Arctic char, Atlantic herring – any will suffice. Until then, I wait.

Day 60

A magnificent day! My shrimp and I (now in my third-stage larval form!) have been gobbled up by an Arctic char! Fortunately as a third-stage larva my tough exterior protects me from being digested by my new host! I celebrate my luck in being another step closer to my goal. The shrimp is not so lucky!

Day 64

I have attached myself to the liver of the Arctic char but I don't cause him any trouble. He is merely a non-essential intermediate host who will ultimately transport me to my next host. I hope this will be a marine mammal so I can complete my life cycle! But what if it isn't?

The worst possible scenario is for my Arctic char to be caught by a person. Oh, the dread of this.

Day 66

A close call. My Arctic char and I were nearly ensnared by a fisherman's net, but my agile host got away. Other fish were not as lucky. I wonder how many of my kind were caught along with their fish hosts. An interesting question indeed. People don't do much surveillance research on *Anisakis simplex* in Canada's North. Just how many fish and marine mammal do we infect? And in what numbers? I can only wish that lots of us are here so we can all contribute to our survival as a species.

Day 68

I have received word from the outside. Several of my *Anisakis simplex* comrades were indeed living in the fish netted by the Inuit fishermen whom I evaded just two days ago. My comrades met an unfortunate end. Their host fish were caught by humans for food! An utter tragedy to my kinfolk.

As third-stage larvae, we are visible to the human eye, often coiled up and penetrating the gut wall or gut organs, but reaching two centimetres in length when stretched. If my comrades were found in the fish, they would surely be picked out and discarded. Lives destroyed, Missions incomplete.

If the fishermen don't see my larvae comrades, they may migrate into the meat of the fish after the fish dies. Cooking would kill my captured comrades. However, like many Inuit communities, fish in this community are often consumed rare, or they are smoked or fermented, processes that do not necessarily kill my kind. In this instance, the humans would eat the live larvae as they eat the fish meat. But alas, this still marks the end of the Mission for these larvae. Humans are nothing to us *Anisakis simplex*. They're dead-end hosts, completely useless. Humans cannot help us in our life cycle. We cannot grow, lay eggs or be passed on from the dreaded human body. To be consumed by a human ultimately means failure in our Mission.

As my host Arctic Char swims in the frigid ocean, I dread this thought, this demise. Sad prospects indeed. I must remain strong. But it further renews my motivation for achieving my Mission for my kind.

Day 77

More word from the outside on my fallen comrades. Some larvae were, in fact, eaten by the humans with the raw fish. A 40-year old woman became ill, with awful stomach pains for three days. She no doubt has what is named after us - "anisakidosis." The disease is most frequent in communities where humans traditionally eat raw or undercooked (infected) fish, such as this Inuit community or in Japan where sushi is common.

Anisakidosis probably isn't a pleasant experience for humans. Sometimes, we larvae don't hang on in the intestine, so we may be coughed up, vomited or come out in the stool. Not glamorous for the larvae or the patient.

Other times, we larvae may burrow into the lining of the human's stomach or intestine. Then they have nausea, vomiting, diarrhea and severe stomach pains. This can last from several weeks to up to two years, and is sometimes confused with stomach ulcers, appendicitis and Crohn's disease. Although rare, I've also heard that some humans have allergic reactions to us larvae, and may get skin welts or even die from anaphylactic shock.

But I've become side-tracked in human-related matters! I must return my focus to my task at hand. Fortunately, I am still inside this Arctic char, hoping to soon be consumed by a marine mammal – a dolphin, a porpoise, a beluga whale – again, I am not particular as to which of the species.

Day 79

I seem unable to keep from discussing human-related matters. My sources indicate the woman who fell ill from anisakidosis was trans-ported south to receive medical care for her abdominal pain. The physician passed a long tube with a videocamera (gastroscope) down her throat into her stomach and found a two-centimetre long, twisting "worm". They quickly removed it with long grasping forceps and sent it for testing. It was one of our third-stage anisakis larvae. The woman is expected to make a full recovery. The larva is not.

Oh, how I hope a marine mammal will soon consume my Arctic char host and me! My sense of urgency to complete my life cycle is even stronger with this news.

Day 95
I continue to wait within this Arctic char. It would appear my host is quite skilled at evading predators. I can only remain patient.

Day 120
Oh, praise the parasitic powers that be! A beluga whale ate my Arctic char and me this morning! Inside this new host, I can now begin the final stage of my life cycle!

Day 121
I'm happy to report that I'm not alone in my beluga. Dozens of other *Anisakis simplex* are here, each in the process of maturing to adulthood. You could say we are all teenagers hanging out inside the whale. This can only mean good things for my finding a mate! Our numbers strengthen my confidence in our Mission, and the preservation of our species.

Day 136
I remain in the beluga and continue growing into an adult.

Day 180
It has been a long, glorious journey. Since making it into the beluga, I have completed two moultings and have matured into an adult anisakis worm. I am now longer, thicker and sturdier so I can deal with the harsh environment in this marine mammal's intestine. I have fed, mated and laid hundreds of thousands of eggs. Soon, my beluga host will shed my eggs in its feces in the cold Arctic water, starting them on their own journeys, and my Ultimate Mission will be complete.

Day 182
Alas, my time in this world has come to an end. Today, Inuit hunters caught my beluga, which is a tremendous event for the Inuit community and a great source of food for its people. Sadly, this event also marks the end of my host and my journey. Fortunately, the beluga had already shed my eggs into the water, so I can depart knowing my Mission was fulfilled.

Day 183
A strange turn of events. I seem to be in the laboratory of researchers investigating *Anisakis simplex* in this community. These researchers have set up a nearby lab and are working with the Inuit

fisherman to determine what fish and mammal species we infect and in what numbers.

The hunters shared my beluga's stomach and its contents with the researchers, who will now look for us. As adult anisakis, we aren't a health concern for people; it is only our third-stage larvae that can infect them. But they are likely interested in us because we produce the eggs that continue the anisakis life cycle.

What does this research mean for my kind? Are we threatened if humans know more about us? I suspect not. Our complex life cycle involves multiple species living in a large aquatic ecosystem - that should thwart any direct manoeuvres against us!

It's more likely the humans will work towards reducing disease in their own people, probably through education. For instance, they will be encouraged to gut fish immediately after capture to prevent the larvae from migrating from the gut cavity to the meat of the fish. Alternatively to kill anisakis, they can cook fish to greater than 60°C for at least one minute or freeze the fish to minus 20°C or below for 24 hours.

But humans are an odd bunch. Their cultural practices, personal preferences and habits aren't always easily changed.

In the end, these efforts make no difference to us anisakis as a species. If people catch us in our animal hosts, our fate is sealed. What's this now? The beluga stomach is being opened. There is a pair of forceps approaching, and liquid-filled jars, and I…

Parasite's Log Ends.

Figure 20 Anasakidosis

The Old Man's Guilt

Donna Curtin

If only I had known I was going to make the little girl sick, I would have left long ago. I'm not nasty by nature, though it may seem that way to my family. I once considered myself quite handsome, but time has expanded the waistline, thinned the hair, and the muscles that in my youth professed to my strength, have long since gone. I live with my Bobby and his wife, Sylvia, though I prefer not to consider her my in-law. Bobby has features very similar to my own, dark-set eyes and a well-fed pudgy midline, while his waif of a wife, Sylvia, is stingy on all that the good-life has to offer. I sit in this box of a room all day and observe their life as though they are my television to reality.

I can't deny that I was bitterly resentful the day my Bobby brought Sylvia home; and then the woman never left. We were stuck with her. Well, at least I was stuck with her. The boy escapes to work every day and I'm left at home with the clean freak. My general untidiness never bothered my Bobby before she came to live with us. I used to have free reign of this house and now I'm confined to the back porch. I'll admit that I'm messy - but after the wedding and the arrival of the baby girl there was no living with her. She probably wants me to move out. But am I supposed to just wander off into the wilds to fend for myself after all these years of suburban captivity? The world has changed so much and besides, all my friends are dead. Who would I know after all these years?

I've been half-expecting death by vacuum ever since Sylvia cradled the little bundle of screams into the house. I used to plot ways to silence the baby, until completely against my will and better judgement, she pick-pocketed my affection with those glittering dark brown eyes and bouncing blond curls. Now I uncharacteristically skip my morning nap, just to listen to her giggle as I roll Cheerios onto the floor after her mother has finished sweeping. I was a firm believer in the adage that children should be seen and not heard before my world shifted for little Magdalene. It puts Sylvia over the top when Mag comes to sit by my feet. I snort and hiss and make faces that throw her into fits of laughter. Sylvia's just jealous because she never takes the time to slow down and enjoy the simple pleasures. With her it's all 'put your shoes on the right feet', 'brush your teeth', 'take the Tupperware off your head', so serious, never any fun.

About six months ago, we all knew something was terribly wrong. I was minding my own business, dipping apple slices in my water because I never trusted that woman to wash things thoroughly for me, when I heard an unusual sound. Sylvia was in the kitchen, wiping down the counters for the millionth time while three- year-old Mag was loading stuffed bears into a cavity she had created by stacking couch cushions. I heard a loud, solid thud and looked up. By the sheer power of the thump I assumed that Mag had hit her head or fallen from a height. I anticipated a holler, a scream for 'mommy', but none came. She must have winded herself, I assumed. I was waiting for the cry to break through the breathlessness, but it didn't and this wasn't right. Then there was only the deafening shuffles of this little body, no more than forty pounds thrashing on the floor.

Sylvia came in, starting immediately into a lecture about the mess of toys on the sofa and then she saw Mag. Twitching, Mag stared at the ceiling, her blond hair in disarray across her forehead, tiny toes flickering across an imaginary keyboard. I sat back in shock as Sylvia gasped, dove to the floor and scooped tiny Mag into her arms. It couldn't have lasted longer than a minute or two, but it felt like forever. Eventually the miniature body ceased to tremble and a dark spot spread across the little girl's pink tights. She must have wet her pants. Sylvia spoke in such a voice as I have never heard before. She was so hushed, yet urgent: "Magdalene, Magdalene, are you okay? Speak to me baby." She kept methodically wiping Mag's hair back off her face, feeling her forehead and at the same time, soothing herself. Mag's flaccid arms flopped across Sylvia's back every time Sylvia would helplessly bear hug the lifeless child.

Then I saw the small hand start to move, no longer twitching. The deliberate motion extended to Mag's wrist and then her elbow, until she wrapped her little arms around her mother. I let out my breath and swallowed hard as I saw Mag start to coil a lock of Sylvia's hair around her forefinger. Mag loves to twirl hair; she started as a baby, first her own hair, then Sylvia's, sometimes even Bobby's or mine when she's really tired. Sylvia began to cry and weep with joy as her daughter regained consciousness and in that moment, I liked that woman. She reached for the phone and dialled 911.

I was nowhere near precious Mag that horrible day, so I had no inkling that I was at fault. Even Sylvia, who likes to blame me for every mess, every smell, never thought to connect me to Mag's illness - until the day the veterinarian came for a house call. About two months after that first seizure, things were getting back to usual, but Mag continued to have seizures, creating an air of urgency and frantic fear about the house. Sylvia was overwhelmed with Mag's unending medical appointments. And she was more militant than ever; my slovenly ways a constant irritation to her. She couldn't dictate the paediatrician's appointment schedule, couldn't control the outcome of the MRI's, or prevent the needles from being stuck into her baby, but she could force me to live in confinement.

Bobby knew better than to ask his wife to stuff the cat into a box and drag the fat old retriever into the car for their annual trip to the clinic, so he asked the veterinarian to make a rare house call. The veterinarian sat at the kitchen table that morning, black bag folded out, stethoscope draped around her neck, medical records dated and signed. The veterinarian started with Sylvia's horrible calico cat, Calypso. Why would you keep a scrawny animal as miserable as that? She hisses at me constantly and ejects fur balls like a gumball machine. Sylvia complains about *my* messes, but Calypso is disgusting. The veterinarian skilfully kept the cat from launching off the table while she inserted the rectal thermometer, much to my vengeful pleasure. They began talking and I gathered she had been good friends with Sylvia in high school.

"I've never been more terrified in all my life, Amy," Sylvia said, "I'd never seen a seizure before; it was like her little body had been possessed."

"It's scary enough to watch a client's pet have a seizure; I can't imagine my own child. Do they know what's causing them?" asked Amy.

"No, the worst possibility would be brain cancer, the best, perhaps epilepsy."

"Sheesh."

"Yeah. They keep scrutinizing our life, asking into everything; chemicals, toxins, poisons, travel, pets . . . they have so many questions. I'm never sure what they are getting at. I keep expecting the Children's Aid Society to show up on the doorstep. I feel so guilty, as if they think we are beating her into these seizures."

"I'm sure that isn't the case, Sylvia. They have to be thorough," said Amy. Then, after a short pause, she said, "They asked about your pets?"

The veterinarian looked about the room, glancing over the fat retriever, "Dodger" in his usual sluggish state, beside his neat wicker basket full of squeaky toys. Then her gaze came upon me and intensified; her brow creasing. I was startled, almost jumped at the attention and let out a snort. I felt instantly uncomfortably, like an object. I'd been living with my Bobby longer than anyone and yet I suddenly felt like an intruder.

"How long has he been with you?"

They were speaking as if I couldn't hear them, as if I couldn't understand.

"Oh, him," Sylvia spoke with her usual derision, "Bobby brought him home when his mom died. His mom took such good care of him that Bobby figured he'd never survive on his own. Bobby and he lived together like two bachelors for years before Bobby and I started dating."

The wrinkle on the veterinarian's forehead deepened. She kept staring at me, thinking. I shuffled under the scrutiny, spilling my cereal, causing Sylvia to puff out a sigh.

"Does Magdalene play with him?"

"She loves him. She sits for hours at his feet, just watching him make faces. Why?" Sylvia, too, was getting uncomfortable as the veterinarian's attention stayed focused on me.

"I'm not sure, but you may want to look into whether he could have something to do with Magdalene's illness."

That was it. I didn't hear another word. All I could see was that accusing look. My fault? How on earth could it be my fault? The veterinarian carried on with her examinations, patted down, poked and prodded each pet, peered into all of their orifices and then she needled them. She produced something for flea prevention and discussed cleaning the cat's teeth. All the while my world was crashing around me. The veterinarian left with the assurance that she would do some research and return with more information about a possible cause of Magdalene's seizures, a cause that had something to do with me.

That night when Bobby came in from work, Sylvia pounced on him before his keys had even jangled into the dish on the side table.

"Amy thinks that," she sighted a kill shot right into me, "he could have something to do with Magdalene's seizures!"

Bobby shook his head in confusion and then he let out a breath, visibly trying to hold onto his temper, saying "let's start again." He mocked a cheerful tone. "Hello dear, great that you're home. Why, yes, I did have a long day. Now what the hell are you saying? Amy who?"

"Amy, our veterinarian, my friend from school. She came to vaccinate Dodger and Calypso today. Do you never look at the calendar? I swear you think that all that I do is watch Oprah all day."

"Back up, say this again. She thinks who could have something to do with Mag's seizures?"

I hid my face in my hands; he had just used my pet name for Magdalene and that would only escalate Sylvia more.

Sylvia let out an exasperated breath, then she pointed directly at me, "Him, she thinks him and his disgusting messes could have something to do with our Magdalene's seizures. They've been looking for brain cancer for God's sake, brain cancer. If he's the cause of this, so help me . . . ."

"Wait a minute, just wait a minute. What on earth are you talking about?"

"I'm not exactly sure, Amy said she would come back later tonight with more information, but I guess there is this parasite that can cause serious illness in small children."

Again, the world went numb. Again, I felt like an intruder, a voyeur into their happy little domestic life. All I do is eat and sleep and plot ways to make little Mag laugh. I would never want to harm her.

During a very strained dinner, my Bobby and Sylvia spoke mostly to Mag and not to each other, while completely ignoring me, as if I hadn't heard their fight. Then, Amy returned.

"This is a very rare disease, let me assure you of that first. But he lives with you, and Magdalene has close contact with him, so you should at least give this information to your paediatrician. I can check to see if he even has the parasite and that may help."

"Amy, can you explain this to me again," Bobby asked dejectedly.

The veterinarian turned to look at me, "Raccoons can carry a round worm called *Baylisascaris procyonis*. We generally just call it 'raccoon roundworm'. It's a typical round worm, similar to what your dog or cat would carry, except that it's found mainly in raccoons. A recent discovery showed that dogs can pick up this worm and spread it as well. All the more reason to de-worm your pets." She gestured to Dodger, who ignored her by snoring away.

"We've kept your cat and dog on parasite control especially after Magdalene was born. If a small child ingests a dog or cat roundworm egg, the egg can hatch and the larva can migrate to the back of the eye or under the skin. It's a rare disease called 'visceral larval migrans'; to be safe, we de-worm pets routinely."

The veterinarian shifted her attention back to my Bobby and Sylvia, who were listening intently. "I only knew about your dog and cat because they come into my office. When you mentioned the seizures, I vaguely remembered something from veterinary school. Living and working in the city, I don't often see pet raccoons, so I had to do some research."

My Bobby slowly shifted his eyes over to me. He'd never looked at me that way.

"So, you think this raccoon roundworm could have something to do with Magdalene's seizures?" Bobby asked.

"It's possible. A very remote chance, but possible. You need to let the doctors know that you have a pet raccoon and that your little girl loves to play with him. If the raccoon has this roundworm, Magdalene could have picked it up when she crawls under his cage and then put her hand in her mouth."

Sylvia gasped, and then walked to the door that leads out to the deck where I live and forcefully shut the door. I could still hear them, because the window beside my cage was open.

"Raccoons with baylisascaris roundworms can deposit thousands of eggs into their latrine, their bathroom area. One adult worm alone can produce over 150,000 eggs per day," said Amy. "These eggs are very strong and hardy. No matter how much Sylvia cleans, there is no way that she could remove all the eggs. I'm sure you've noticed that racoons are incredibly messy. Do you know why your mother kept him?"

"She found him in her garage attic, chattering and chirping, after the neighbour live-trapped a large female raccoon and took her away," Bobby replied. "I remember feeding him pieces of banana when he was so small that he fit into my palm. He was so adorable. Now look at him, he's got to be more than 20 pounds."

The veterinarian went on to explain, "I can run a fecal on him to look for the roundworm eggs. You could ask your doctor if Magdalene should have an anthelmintic, a de-wormer."

"What does this raccoon roundworm do to children?" Sylvia asked impatiently.

"It's hard to say exactly because it's difficult to diagnose. Unfortunately, the diagnosis is usually made when they find the parasite in the brain of a patient who has died with a neurological disease such as blindness or seizures."

Sylvia looked about the room, swiftly moved across the floor, picked up anything out of place and returned it to order, controlling what she could. Then she noticed my window and firmly slid it shut before she finally came to a standstill by my Bobby and cupped her elbows in her hands. I'd never liked the woman because she's always so short with Bobby. But I have to admit that she's been a good wife to him and a dedicated mother to Mag. Even through my guilt and now isolation, I felt sorry for her.

Bobby reached out and pulled Sylvia to his side, leaned over and kissed the top of her head. He didn't look at my cage, and I wondered if he ever would again.

The veterinarian took a small bag from her pocket and came out on the porch. She stared down at me. She took in my striped tail, round torso with skinny legs, pointed snout and curious keen eyes. I sniffed the anti-bacterial cleanser on her hands with my strong sense of smell. The chemical odour made me snort, which caught the veterinarian off guard. She jumped and I let out a startled hiss.

She laughed at her jitters. Then she knelt down and with the bag over her hand she scooped up a small amount of droppings from under my cage. She folded the bag back over itself and tied a neat knot to close it.

I knew I would have to leave. For the first time since I'd come to live with Bobby, I felt like a colony of disease-infested vermin. For days I studied my cage for weaknesses; scrutinizing the rusty hinge corners of the hatched wire walls, mentally untwisting the wire that looped through the lock. Sometimes I could sense someone watching me. Standing inside the house, her barefoot little body barely sinking into the plush living room rug, her princess nightgown a pink parade, Mag's brown eyes framed by dark long lashes looked into mine. She was as adorable as ever, but there was a shadowing under her eyes that I had never seen. It had been a bad night for the family. Mag had a long seizure during the night, which had kept her up for hours and her parents for much longer.

We used to spend our mornings together, Mag sharing her cereal, while I annoyed her mother. I would reach through the mesh of my cage and hold my hand open; she would reach out with one of those tasty cereal pieces and after repeating this process several times, she would offer me her finger. I would clasp her chubby soft finger and just hold it. I never pulled, I never scratched or hissed or scared her. We would just sit for a moment or two, in our own way, holding hands. But I know that I will never get to touch little Mag, as this routine ended after the veterinarian came to collect my feces sample.

The door to the porch has been locked. Mag tried to open the door that first day, pulling at the handle, hanging from it, screaming to melt into a fit of frustration and anger. I felt so helpless.

Now Mag doesn't even try the door. She just leans into the window, placing her palm down onto the pane of glass to showcase the tiny weave of a lifeline that has barely had the chance to tell her future. Then she lifts up her other hand with a secret offering. She smiles and holds out to the glass pane a single mushy Cheerio, her hand not much bigger than my own.

***

Sylvia comes back through the living room to grab the file folder with all her notes, research and questions for the neurologist. Magdalene is settled into her car seat in the garage, her portable DVD player showcasing Nemo's adventures. Sylvia turns to leave when she feels that something is off. Is she forgetting something? She looks around the room, mentally checking off her morning; the milk is put away in the fridge, Calypso and Dodger have been fed, the stove is off. Then she sees the empty raccoon cage. He could be in his box house, but he usually sleeps out on the cage floor during the day. Sylvia peaks into the garage to check that Mag is still content in her car seat, then she carefully enters the screened in porch. The cage door is open; the wire that once twisted through the lock is unwound and lying on the deck below. A large hole is torn through the screen door that leads into the back yard.

Sylvia goes out and scans the back yard. The large male raccoon is standing on all fours, his spindly legs supporting the large round body. Hearing her open the door, he doesn't move on; he turns at an angle to stare at her, a dark mask with shiny black eyes. There is age and perhaps wisdom in those dark eyes, even though his face has not greyed with time. He takes her in for a long moment. It is almost like he wants to say something, Sylvia thinks foolishly. Then he begins to waddle off, picking up speed as his large body undulates awkwardly, yet oddly gracefully across the lawn. When he hits the edge of the lawn, where the trees and shrubs of the small forest lot behind their home begin, he hesitates for the slightest moment. And then he's gone.

Sylvia is stung then by a small pang of guilt. Should she have tried to call him back? Should she go after him or phone Bobby so that he can return home to find him? But she won't. She will let Bobby discover the empty cage when he gets home from work and very little will be said. The raccoon's leaving is for the best, even if he isn't the cause of Magdalene's seizures. Amy called the day after she took the stool sample to report that the fecal test was negative; no evidence of the dreaded roundworm. But just like Amy said - raccoons don't make good pets.

The neurologist assured them that with a negative fecal test, racoon roundworm disease was very unlikely, but they treated Magdalene just in case. Finally after all the MRIs and other tests, they decided that Magdalene has juvenile epilepsy and started her on medication. The medicine has helped. There would still be rough nights, but the seizures were already coming less often and they weren't as scary. The appointment with the neurologist this morning would hopefully answer more questions and then perhaps they could start to put their lives back together.

The whole roundworm panic had been a false alarm. Nonetheless, after the scare even Bobby wasn't at ease with the raccoon. He just couldn't make the decision to get rid of him. Somehow, Sylvia thought, the animal had done them a favour by leaving. Tonight after dinner the first order of business would be to clean away that smelly cage.

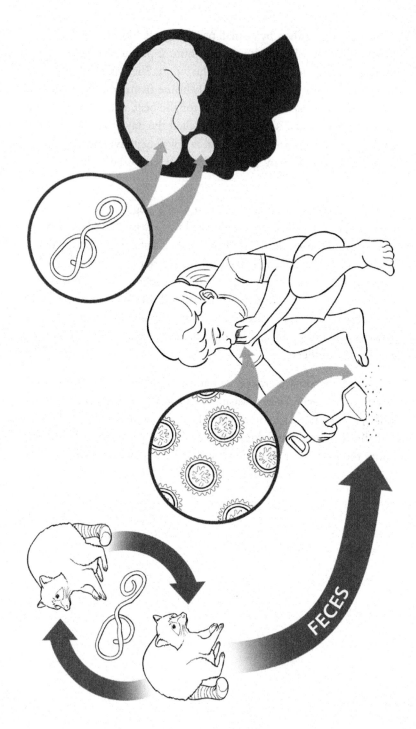

Figure 21 Raccoon Roundworm

A Mite-y Mystery!

Trace MacKay

I received the call just after noon as I was finishing my morning medical records and calling clients about their pets that I had seen earlier in the week. I was paged to the office I shared with three other veterinarians in a busy companion animal hospital in Orlando, Florida.

"Is Doctor T there?" asked the booking receptionist, with a smirk. "Can you let her know I booked another quirky person for her this afternoon? It's a third opinion. The medical records are being faxed over."

Chuckles quickly spread throughout our small closet of an office as I plunked my head into my hands, letting a raspy sigh escape, and thinking here we go again.

Many veterinarians take special interest in certain body systems, ailments or conditions, such as conditions of the heart, lumps or bumps, or cats that pee in strange places. These special cases are directed towards the appropriate veterinarian in any given hospital or town. These unofficial referrals make both the veterinarian and client happy. I tend to attract the eccentric types wherever I work. I had only been in Florida for a few months and had already gained a reputation at our hospital, and seemingly throughout all of central Florida, as the veterinarian for eccentrics, hypochondriacs, the chronically misunderstood and otherwise "demanding" people.

My reputation for these types of recommendations started in my final year of veterinary college. Once, while assigning cases during morning rounds on my small animal surgical rotation, a resident goaded me by saying tongue in cheek, "Trace, you like people who are a little odd. Can you go to room 5 and speak with Ms. X? She is here to discuss treatment options for her dog's mandible cancer. Oh, did I mention she is disappointed the mistletoe extract she made is not working and that she has to leave before 5 pm because the mother ship is landing tonight and her welcome sign isn't finished yet? Good luck!"

I don't remember broadcasting to the world my preference for "unusual". My colleagues assumed that with my sense of humour and my own background I have a greater acceptance of eccentricities – and the designation stuck to me.

I made my way to the receptionist station to find out more about my next appointment. The staff greeted me with knowing grins and a stack of faxed medical records: three cats worth to be exact. When the client made the appointment, she informed the receptionist that she wanted treatment for scabies because the other veterinarians didn't know what they were doing. No pressure on me! I had to figure out why she wanted her cats treated for scabies (a mange caused by the mite, *Sarcoptes scabiei*) and how she had lost faith in the other veterinarians. Glancing at the stack of faxed records as I made my way back to the office, I noticed the bold black capital letters on the coversheet from the last veterinarian, stating "Good luck with this one!" This was going to be a long afternoon.

I learned from the records that two months ago the client was concerned about fleas when she took her cats to their veterinarian. The cats scratched a lot and the client, her husband and two children had itchy, red marks on their skin. The veterinarian didn't find fleas, but did treat and prescribe flea control for the cats. A week later, the client wanted her money back because the cats and the family were still scratching. The veterinarian encouraged the client to have her house treated for fleas and offered a courtesy recheck appointment. A few weeks later, the owner client complained the 'flea problem' hadn't been controlled and wanted her records faxed to another veterinarian for a second opinion.

The second veterinarian examined the cats and could find no evidence of skin problems or visible signs of fleas or other parasites. The owner lifted her shirt to show the veterinarian the lesions she had on her back and sides. The veterinarian noted tiny red bumps that could have come from flea or mite bites. The veterinarian took samples from each cat by scraping the skin surface with a scalpel blade and then examined the skin debris under a microscope for mange mites. None were found. The veterinarian recommended skin biopsies, but the owner didn't like the cost of the biopsies and wanted the cats treated for mange. The veterinarian refused to do this treatment without a specific diagnosis and suggested the owner and her family go to their own physician. She could also contact a pest management company. The last entry in her records appeared a few weeks later and indicated the records were being faxed to our hospital.

I took a moment to summarize what I knew at this point: Three cats who had been scratching for 2 months for no obvious reason and four people in one household with small red itchy spots of unknown origin. I couldn't tell if the house had been treated or the people had been seen by a physician.

As I began to generate a list of questions, I realized I had very few pieces to the puzzle. I was starting to doubt if I was the right veterinarian to solve this case. What if this was some tropical disease or parasite with which I was unfamiliar? During my six months in Florida, I had already seen more ringworm, hookworm and heartworm than most Canadian veterinarians see in a lifetime. I tried to convince an associate who had trained in Florida that she was better suited for the job. Despite my promise to bring her back a family-sized bag of ketchup-flavoured potato chips the next time I ventured home to Ontario, she begged off seeing what was now 'my' case!

I couldn't get ahold of the two previous veterinarians in the short time before the appointment. The receptionist for one of the other veterinarian let me know that she thought the owner was a real problem. Great - this was going to be a very challenging encounter!

Soon an itchy, agitated client and three cats arrived. The client quickly informed the technician she was there for one reason: to get medication to treat her cats for mange mites. She didn't want any more expensive tests and the veterinarian had better know what she was doing.

With a chorus of "good luck" from the other veterinarians, I went to the exam room, hoping I at least appeared confident and competent. Before entering, I peeked in the window; I didn't want to make a bad first impression by opening the door too quickly and flinging one of the cats across the room. I saw three healthy looking cats exploring the exam room and one normal looking woman, not much older than me, staring at the floor and scratching her arm.

As we shook hands, the woman introduced each cat by name and commented on my accent, inquiring, "Are you from 'up north'?" I answered that I was from as "up north" as you can get: Canada! I may have seen a flutter of doubt cross her face, but I could be mistaken. The owner was scratching a healing wound on her forearm and I asked her if that was from a bite. "No", she explained, "It's where the dermatologist took a biopsy a month ago". No specific diagnosis was made, they suggested it might be an allergy or insect bite. The owner and her family were treating their rashes with a cortisone cream and taking oral antihistamines, which both seemed to help. I asked her for her doctor's contact information so I could talk with them about a possible zoonotic disease shared by the family and the cats.

I listened to her detailed history of the itchiness problem with her family and cats, interrupting as needed with questions. After 20 minutes I had learned: All the cats lived indoors; no dogs lived or visited her home, no visitors to their home had complained of itchy skin lesions; the owners hadn't made any changes to the house recently; a pest control company had inspected and treated the house twice and had found no evidence of fleas, bed bugs, ticks, rodents or other pests. She also told me her youngest child had more bite marks and itchiness than the rest of the family. I carefully examined and combed each of the cats for fleas, but found nothing wrong – and no fleas. I asked if I could shave a few spots on the cats to look for skin lesions. The owner replied that the first veterinarian had done this and found small pinpoint red raised bumps but no fleas. She also remarked that the second veterinarian had found no such marks and the cats were not as itchy after the monthly flea treatments. Valuable information, but I was no closer to a diagnosis.

Even though someone probably had already asked about any other pets in the house, I had to find out for certain. The owners off-handed response? "Just the gerbil".

No one had asked about it before and the client hadn't mentioned the gerbil to the other veterinarians. The gerbil lived in a cage in her youngest child's bedroom. Before the house was treated for pests, they always made sure to remove the gerbil and its cage from their house.

I asked the client to take her cats home and bring back the gerbil. Although no one paid much attention to the gerbil, she didn't think it was sick. But the client said she trusted me and believed I was going to solve this two-month-old mystery.

While calming down our receptionist who had just witnessed my client breeze past her desk without paying, I started to panic about how little I knew about gerbils. As a child, I had an unintentional gerbil colony thanks to my best friend who thought she gave me two female gerbils for my 10th birthday. I had learned about small rodents in veterinary school, but didn't remember any weird external parasites or mange. I was no gerbil expert.

The owner returned, carrying "Snarls" in a small, white pastry box for a makeshift gerbil carrier. In a typical pet-naming fashion, Snarls was really quite pleasant. (It's often the 'Sweeties' and 'Princesses' of the world that need to be watched!) Looking inside the pastry box, I was startled to see lots of small, red spider-like critters teeming all over the box and bedding. The owner said she first saw them when the gerbil started running frantically back and forth inside the makeshift carrier. I plucked out a few of them to examine under the microscope. Meanwhile, the technician got rid of the pastry box and bathed and dried 'Snarls'. Looking at them under the microscope, the suspects were diffusely red, about one mm long, eight legs and had false heads with what appeared to be biting mouth parts. They looked like the red spider mites I'd seen as a child on sidewalks, under trees or with nesting birds. I needed to find out what kind of mites they were and how we should treat Snarls. I called my favourite veterinary parasitologist, Dr. Andrew Peregrine, at the Ontario Veterinary College.

I was in luck. Andrew was in his office and with his vast amounts of knowledge of all things parasitic, he quickly determined that they were bird mites, most likely the red poultry mite, *Dermanyssus gallinae*. When birds aren't around, these mites seek out rodents and even will take blood from any warm-blooded animal, including humans.

The owners didn't have chickens or pet birds. Where did the mites come from? Typically, the mites could come from a bird nest in a chimney, vent or window air-conditioning unit. During the day, the bird mites live in the bird nest and at night became active, running around and biting the people in the house. Their daytime disappearance and drive-through fast food feeding habits make these mites hard to detect. The bites on humans cause itchy bumps, occasionally with bleeding centres, which doctors can confuse with more common parasites, such as fleas or sarcoptic mange mites.

After careful searching, my client found an abandoned bird nest in the chimney. The probable scenario: When the birds deserted the nest, their mites moved to the gerbil cage and at night fed on the family members and the cats. Once the cats had been treated with long-acting flea medication, the mites only targeted the human family members. The mites don't stay on their hosts; if the owner hadn't picked up some bedding from the gerbil cage, we could have missed the diagnosis. As a follow-up, I talked with both veterinarians and the client's dermatologist. In medical journals, I found two case reports and a review about people with bites from avian mites living in gerbil cages. I shared these with my now happy client, who was relieved she had an answer and that her cats didn't need further tests or treatments.

The family had the chimney cleaned and capped, continued to treat the cats and Snarls for parasites, and brought in the pest control company one more time. The gerbil got a fancy new mite-free cage and the family no longer suffered from itchy rashes once the seemingly invisible marauders were no longer assaulting them.

For myself, I added a new zoonotic disease to my list for itchy animals and people. On a routine health check for the cats a few months later, the owner thanked me again for solving the mite-y mystery. As I helped her to her car with the three cat carriers, she stopped and said, "You know, I think people thought I was just crazy or something." I didn't mention the observation of the receptionist of one of the other veterinarians.

This case drove home to me some important reminders: Veterinarians should keep channels of communication open with family doctors and other caregivers. Veterinarians should find out about every animal and family member in the house. And perhaps most importantly, veterinarians should remember that clients who are considered to be 'difficult' might just need someone to listen and ask more questions!

Figure 22 Avian Mites

About the Editors

Elizabeth Arnold Stone, DVM, MS, MPP, DACVS, is the dean of the Ontario Veterinary College, University of Guelph, where she has spearheaded the establishment of the U of G Centre for Public Health and Zoonoses, including a new Masters of Public Health degree program.

She previously was a professor and department head at North Carolina State University and a professor at University of Pennsylvania. She completed her DVM at University of California, Davis; surgical residency and MS at University of Georgia; and Masters of Public Policy at Duke University.

Throughout her career she has worked and written on the role of veterinarians in society and on student learning and engagement. She is co-editor of the book, *Animal Companions, Animal Doctors, Animal People*.

Cate Dewey, DVM, MSc, PhD is professor of epidemiology, ecosystem approaches to health and swine health management and the Chair of the Department of Population Medicine at the Ontario Veterinary College, University of Guelph. Dr. Dewey continues her work in the developing world through ongoing education to the people working in pig keeping industries and by expanding the outreach to AIDS orphans. To find out more about the Children of Bukati, please visit www.childrenofbukati.com.

About the Contributors

Don Barnum

The late Don Barnum, DVM, DVPH, DVSc, was professor emeritus at the Ontario Veterinary College at the University of Guelph. During four decades as a professor, he also served as department chair in the areas of bacteriology, microbiology and immunology. He was the author or co-author of over 100 research papers, was a fellow of the American Academy of Microbiology, councillor of the Canadian Veterinary Medical Association, and president of the Canadian Society of Microbiologists. He is recognized as a pioneer of modern microbiology in Canada and globally in the area of mastitis, *E. coli*, and antibiotic resistance. Working with UNESCO in the 1970s and 1980s, he conducted workshops on diagnostic microbiology in Malaysia, Sri Lanka and Tanzania, and taught veterinary bacteriology in China, a joint project of CIDA and the University of Guelph.

Lea Berrang Ford

Lea Berrang Ford, BSc, MSc, PhD, is an Assistant Professor in Health Geography at McGill University. Dr. Berrang Ford is both a geographer and an epidemiologist, with a masters in Environmental Change from Oxford, and a PhD in Epidemiology from the Ontario Veterinary College, University of Guelph. Her research focuses on spatial health analysis of infectious disease and environmental change, with expertise in Ugandan sleeping sickness. Her research projects include climate change and health adaptation in remote Indigenous communities in Peru, the Canadian Arctic, and Uganda; spatial epidemiology of vector-borne diseases distributions and environmental determinants in Peru and Canada; sleeping sickness mapping and risk modelling; and climate change and health policy. Prior to joining McGill, Lea worked with the Public Health Agency of Canada in Saint Hyacinthe, Quebec as an environmental epidemiologist and medical geographer, specializing in spatial health analysis, vector-borne zoonotic disease mapping and environmental health research.

Edward B. Breitschwerdt

Edward B. Breitschwerdt, DVM, DACVIM, is professor of medicine and infectious diseases at North Carolina State University, College of Veterinary Medicine (NCSU-CVM). He is also an adjunct professor of medicine at Duke University Medical Center and a Diplomate, American College of Veterinary Internal Medicine (ACVIM). Dr. Breitschwerdt directs the Intracellular Pathogens ReMarearch Laboratory in the Center for Comparative Medicine and Translational Research at NCSU. He also co-directs the Vector Borne Diseases Diagnostic Laboratory and is the director of the NCSU-CVM Biosafety Level 3 Laboratory. Dr. Breitschwerdt's research group has published more than 300 manuscripts in peer-reviewed scientific journals.

Heather Bryan

Heather Bryan is a fresh PhD graduate from the Faculty of Veterinary Medicine, University of Calgary, Calgary, Alberta, and is now a Hakai Postdoctoral Scholar in the Applied Conservation Science lab at the University of Victoria, Victoria, BC. Her current research seeks to elucidate the physiological mechanisms by which wildlife respond to environmental change. One of Heather's many passions is to share her enthusiasm for ecology through workshops designed to inspire youth about science and conservation. Heather is honoured to have been selected as an ambassador for Wings Worldquest, an organization that celebrates exceptional women explorers in science.

Dominique Charron

Dominique Charron, DVM, PhD, is Program Leader, Ecosystems and Human Health, International Development Research Centre (IDRC), Ottawa Canada. This story was inspired by the work of dedicated Southeast Asian researchers (the Asian Partnership for Emerging Infectious Diseases Research, www.APEIResearch.net, supported by IDRC). The story is fictional and the names are invented. However, the events are based on real research and the results are now published or in press. Veterinarians Without Borders, (Vétérinaires Sans Frontières – Canada) is involved in this research, and the character of Dr. Kathy is inspired by a Canadian veterinarian.

Jack Cote

Jack Cote, DVM, was born in Guelph and graduated from the Ontario Veterinary College (OVC) in 1951. After a short stint in practice, he returned to the OVC and in 1953 helped establish the college's farm services (ambulatory) clinic. In 1960, he set up the first dairy herd health management program in Canada. He served as director of District 12 of the American Association of Bovine Practitioners and was the organization's president in 1981-82. In 1980 the Ontario Veterinary Association named him veterinarian of the year. After retiring from the OVC in 1986, he worked with the Ontario Ministry of Agriculture and Food as a cattle health consultant. For almost 30 years, beginning in 1970, he raised and raced standard-bred horses as a hobby.

Donna Curtin

Donna Curtin, DVM, practices veterinary medicine in Bruce Country, Ontario, close to her rural hobby farm where she lives with her husband and two children. A mixed-animal practitioner, she obtains career satisfaction through improving the health of the family pet, constantly acquiring new skills and relishing every emergency from late night C-sections to lacerated horses. As a complement to her veterinary career, she aspires to become a published novelist. She has published in the Canadian Veterinary Journal, writes veterinary related articles for local newspapers and is currently working on her novel and a series of short stories related to her experience as a veterinarian. Animals, just as often as people, play important characters.

Robert Curtis

Robert Curtis, DVM, MSc, was raised on the family farm near Orangeville, Ontario. He graduated from the Ontario Veterinary College (OVC) in 1961, winning the Andrew Smith Gold Medal as the top student in his class. He later received a Master's of Science degree from the University of Guelph. He worked in the OVC Department of Clinical Studies until 1985, when he joined the new faculty of veterinary medicine at the University of Prince Edward Island as chair of the Department of Health Management. He received numerous honours including the Norden Teaching Award and the Schering Large Animal Award. He was the honorary class president for the OVC class years of 1967, 1973, and 1975. In 1988 he was given the OVC Distinguished Alumnus award given by the OVC Alumni Association.

Brian Evans

Brian Evans, DVM, is a Deputy Director General with the World Organization for Animal Health (OIE) in Paris, France. Previously he was the Chief Veterinary Officer of Canada, being appointed at the inception of the Canadian Food Inspection Agency (CFIA) in 1997. He was the delegate of the Government of Canada to the OIE from 1999 to 2012 and Canada's first Chief Food Safety Officer from 2010-2012. Dr. Evans has a Bachelor of Science in Agriculture, with a major in animal science and minor in genetics, from the Ontario Agriculture College and a Doctor of Veterinary Medicine degree from the Ontario Veterinary College of the University in Guelph.

Jody Gookin

Jody Gookin, DVM, PhD, DACVIM, is an Associate Professor of Small Animal Internal Medicine at North Carolina State University. Her research interest is the host response to gastrointestinal infection. Her laboratory is credited for discovering *Trichomonas foetus* as a common cause of protozoal diarrhea in cats.

Maelle Gouix

Dr. Maëlle Gouix is truly a "global vet." She has volunteered or worked on diverse species, including dogs and cats on Corsica, marine mammals in Alaska, wolves in British Columbia, and muskoxen and caribou in the Northwest Territories. Her current MSc endeavors at the Faculté de médecine vétérinaire de l'Université de Montréal are studying beluga whales and their parasites in the St. Lawrence River and Nunavut.

Blake Graham

The late Blake Graham, DVM, grew up on a family farm in Ontario and entered the Ontario Veterinary College in Guelph, Ontario in 1947. After graduation, he worked for about one year in California before returning to Canada and opening his own practice in Toronto. He left the practice in 1975 for a career in real estate. Dr. Graham's book, *Sow's Ear to a Silk Purse, Anecdotes from the Life of a Veterinarian*, was published in 2004. Dr. Graham established the Blake Graham Fellowship at OVC to help fund graduate students studying zoonoses at the Master's and Doctorate levels.

Michele Guerin

Michele Guerin, DVM, MSc, PhD, is an associate professor in the Department of Population Medicine at the Ontario Veterinary College. Michele received her DVM from University of Guelph in 1993 and was in private practice for 8 years. She returned to Guelph to complete her MSc (2003) and PhD (2007). This story is based on her graduate work done in Iceland. Her current research focuses on production diseases of poultry in Ontario as well as biosecurity.

Carlton Gyles

Carlton Gyles, DVM, MSc, PhD, is Professor Emeritus in the Department of Pathobiology, Ontario Veterinary College (OVC), University of Guelph. Carlton was born in Jamaica and studied veterinary medicine and microbiology at the OVC, followed by studies in the United Kingdom and Denmark. He was a faculty member at the OVC from 1968 to 2005 and is presently editor of The Canadian Veterinary Journal. Much of his research has involved the bacterium *E. coli* and he is fascinated with the complex mechanisms by which bacteria cause disease in humans and animals.

Char Hoyt

After graduating from Emily Carr in 1997, Char Hoyt went on to explore a number of different creative endeavors, becoming an active member of the Vancouver underground art scene. In 2012 she began creating paper mache animal masks and other art objects for various film projects. Char enjoys giving humans the chance to express their inner animal. Michelle McBeth commissioned Char to create a mask of her spirit animal, the Raccoon. The cover image of this book, photographed by Christopher O'Brien, is the result of this collaboration. To see more of Char's masks and artwork, check out www.charhoyt.com and www.facebook.com/CharHoytArt.

Bruce Hunter

The late Bruce Hunter, DVM, MSc, retired from the OVC Department of Pathobiology after a distinguished 30-year career at OVC and became Professor Emeritus. A graduate of the Western College of Veterinary Medicine, he was an expert on diseases of wild birds, poultry and mink. He taught courses in ecosystem and poultry health, published numerous articles and book chapters, and edited three books including *Diseases of Wild Birds*. He championed the integrating of animal, human and environmental health at OVC and globally. He was instrumental in establishing a Canadian Community of Practice in EcoHealth (CoPEH) and a graduate-level course in ecosystem approaches to health involving three Canadian universities (Guelph, British Columbia, and Montreal). He co-led a poultry project in Ghana for Veterinarians Without Borders / Vétérinaires Sans Frontières-Canada and in recognition received the Aeroplan Volunteer of the Year Award.

Jerry Jaax

Jerry Jaax, DVM, ACLAM, is associate vice-president for research compliance and university veterinarian at Kansas State University. He served in the US Army Veterinary Corps for 26 years and was the chief of veterinary medicine and laboratory support.

Nancy Jaax

Nancy Jaax, DVM, ACVP is adjunct professor, diagnostic medicine and pathology, Kansas State University. During her military service career she became an international expert in emerging zoonotic diseases, specifically Marburg and Ebola viruses.

Claire Jardine

Claire Jardine, MSc, DVM, PhD, obtained an MSc in Ecology from the University of British Columbia after completing a BSc in Wildlife Biology at the University of Guelph. Her interest in wildlife diseases led her to return to school and she obtained her DVM in 2001 and her PhD in 2006 from the University of Saskatchewan. Claire is currently an assistant professor in the department of Pathobiology at the Ontario Veterinary College. Her research program is focused on understanding the ecology of zoonotic pathogens in wildlife populations and she is currently involved in studies investigating antimicrobial resistance, salmonella, *Borrelia burgdorferi*, and leptospira in wild mammal populations. Claire teaches 4th-year veterinary student electives in ecosystem and wildlife health and a graduate course in zoonotic diseases.

Andria Jones-Bitton

Andria Jones-Bitton, DVM, PhD, is a veterinarian and epidemiologist who conducts research in food-borne, waterborne, zoonotic and infectious diseases, small ruminants, and the Arctic. She regularly uses both quantitative and qualitative analytical methods to achieve her research objectives. She teaches graduate and undergraduate classes in epidemiology at the University of Guelph. This is her first piece of fiction writing.

Philip Kitala

Philip Kitala, BVM, MSc, PhD, is an Associate Professor in the Department of Public Health, Pharmacology and Toxicology, Faculty of Veterinary Medicine, University of Nairobi, Kenya. He was born and raised in Machakos County, Kenya. He has a BVM, MSc in Veterinary Public Health and a PhD from the University of Nairobi. As part of his PhD program, Philip was enrolled at the Ontario Veterinary College for courses in 1991 and 1994 to '95. He teaches veterinary and post-graduate students and has supervised 18 MSc and 3 PhD students. He has published approximately 30 peer-reviewed and conference papers. Philip is an internationally recognized expert in rabies and has done a number of consultancies for international organizations. He is a registered veterinarian and member of the Kenya Veterinary Association.

Sandra Lefebvre

Sandra Lefebvre, DVM, PhD, is a two-time graduate of the Ontario Veterinary College at the University of Guelph, first as a veterinarian (DVM) and second as an epidemiologist (PhD). After her research with therapy dogs she joined the American Veterinary Medical Association as a scientific editor for the association's journals. Currently, she is working for Banfield, The Pet Hospital, as a medical (epidemiological) advisor and researcher. Dr. Lefebvre shares her home with two cats, Busy and Kanga, which provide a personal source of pet therapy. Her interests continue to include human-animal interactions and the joys and "potential" perils associated with them, particularly from the animals' perspective.

Gregory Lewbart

Gregory Lewbart, MSc, VMD, received a BA in biology from Gettysburg College (1981), a masters in biology majoring in marine biology from Northeastern University (1985), and a VMD from the University of Pennsylvania School of Veterinary Medicine (1988). After working in the private sector, he moved to North Carolina State University, College of Veterinary Medicine (1993), where he is Professor of Aquatic Animal Medicine. Dr. Lewbart has authored over 100 popular and scientific articles about invertebrates, fishes, amphibians and reptiles and speaks locally, nationally and internationally on these subjects. He has also authored or co-authored 20 book chapters related to veterinary medicine on these taxonomic groups and edited or co-edited three veterinary textbooks. Additionally, he has two published novels, *Ivory Hunters* (1996) and *Pavilion Key* (2000). He and his wife, Dr. Diane Deresienski (VMD, DABVP), live in Raleigh with their assorted pets.

Tony Linka

Tony Linka is a Toronto-based artist who graduated with honours in the Bachelor of Arts Scientific/Technical Illustration program at Sheridan College. Currently, he works as a freelance illustrator on a broad range of projects which include illustrations for books and magazines; 3D renders for product visualization; and graphic design for merchandise and exhibit displays. He also works as a Character FX and Lighting Artist in the television and feature film industry.

Trace MacKay

Trace MacKay, DVM, MPH, has come a long way since she was disillusioned at age five after finding out she would never grow up to be a giraffe no matter how hard she tried, and set out to explore alternate career paths. She completed an honours BSc in Biological Science at the University of Guelph (1999) and a DVM degree from Ontario Veterinary College (2003). After serving as a locum and emergency veterinarian, she returned to OVC and completed a master's degree in public health (2010). Dr. MacKay has instructed medical doctors on zoonotic diseases, worked as a research consultant for clients including Ronald McDonald House Children's Charity,

Let's Talk Science and IDEXX Laboratories, managed a Guinea Fowl production project in Northern Ghana for Veterinarians without Borders Canada and currently works as a research and evaluation manager for Community Food Centres Canada (and as a veterinary locum in Toronto in her spare time!).

S. Wayne Martin

S. Wayne Martin, DVM, MPVM, PhD, grew up on an Angus cow-calf farm near Udney, Ontario. Following graduation from the Ontario Veterinary College in 1967, he joined Amherst Veterinary Clinic in Scarborough. In 1969 he returned to OVC to complete an MSc program. Thereafter, he completed the MPVM and PhD (Comparative Pathology - Epidemiology) at University of California, Davis and then joined the OVC department of Veterinary Microbiology and Immunology. In 1987 he became the founding chair of the department of Population Medicine. He was appointed as a Fellow of Canadian Academy of Health Sciences in 2005. After retiring from U of G in 2006, he has continued to consult internationally.

John McDermott

John McDermott, DVM, PhD, graduated from the Ontario Veterinary College (DVM, 1981; PhD, epidemiology, 1990) and joined the Department of Population Medicine. In 2003 he was appointed the Deputy Director General and Director of Research at the International Livestock Research Institute. John and his wife Brigid and daughters Patricia and Rebecca lived and worked in Africa for 25 years. In 2011 he became Director, CGIAR Research Program on Agriculture for Nutrition and Health in Washington, DC. His research focuses on public and animal health, and livestock research in developing countries, primarily Africa. He has led projects on zoonotic and emerging diseases in Asia and Africa. He has authored or co-authored 200 peer-reviewed publications, book chapters and conference papers, and has advised over 30 postgraduate students. The University of Guelph awarded him an Honorary Doctor of Laws Degree in 2012.

Tracey McNamara

Tracey McNamara, DVM, DACVP, is a board-certified veterinary pathologist specializing in diseases of captive and free-ranging wildlife. She is currently a professor of pathology at Western University of Health Sciences, College of Veterinary Medicine in Pomona, California. In 1999, she played a key role in the discovery of the West Nile virus, a virus that had never before been seen in the Western hemisphere. Since that time, she has been involved in integrated biosurveillance projects in the United States, Russia, Uzbekistan and Kazakhstan and continues to promote closing the gap between veterinary medicine and public health. She shared a version of this story in the book *Secret Agents: The Menace of Emerging Infection* (Penguin: 2002).

Paula Menzies

Paula Menzies, DVM, MPVM, is a member of the Ruminant Health Management group in the Department of Population Medicine and is a ruminant field service clinician at the Ontario Veterinary College. She is involved in health management issues for the sheep and goat industries at the provincial and national levels, including flock health management programs, biosecurity, on-farm food safety and specific disease control programs. Her current research activities include not only Q-fever in sheep and goats and their farm workers, but also gastrointestinal parasite control in sheep, Johne's disease in dairy sheep and goats, drug use and antimicrobial resistance in sheep flocks, *Cysticercus ovis* infection in sheep, and scrapie surveillance.

David Pearl

David Pearl, DVM, MSc, PhD, is an associate professor in the Department of Population Medicine at the Ontario Veterinary College. His research focuses on the surveillance and epidemiology of zoonotic and animal diseases. He completed his PhD (2006) and his DVM (2001) at the University of Guelph. Prior to coming to Guelph, he received his MSc (1994) and BSc (1991) in Biology from York University and McGill University, respectively. While studying and working as a biologist, he was involved in a variety of projects related to ethology, ecology and wildlife conservation.

John F. Prescott

John F. Prescott, MA, VetMB, PhD, emigrated from the United Kingdom in 1976 to work as a veterinary bacteriologist at the Ontario Veterinary College. He has diverse interests in bacterial infections in animals, including leptospirosis, but is best known for work on *Rhodococcus equi* pneumonia in foals and in promoting better use of antimicrobial drug use in animals. He is an editor and an author of the textbook *Antimicrobial Therapy in Veterinary Medicine*, now in its fifth edition. His more recent research interests are in enteric diseases of animals caused by type A *Clostridium perfringens*. He was elected a Fellow of the Canadian Academy of Health Sciences in 2008.

Jan M. Sargeant

Jan M. Sargeant, DVM, MSc, PhD, received her DVM degree from the Ontario Veterinary College (OVC) and practiced as a food animal veterinarian for 4 years. She then obtained an MSc and PhD in Epidemiology from the University of Guelph. Jan has been on faculty in the College of Veterinary Medicine, Kansas State University and the Department of Clinical Epidemiology and Biostatistics, McMaster University. Currently, she is Director of the University of Guelph Centre for Public Health and Zoonoses, and a professor in the Department of Population Medicine at the OVC. Jan's current research interests are the epidemiology of zoonotic pathogens, linking research across disciplines and among animal health and human health communities, and evidence-informed decision-making in public health.

Judit Smits

Judit Smits, DVM, MVet Sc, PhD, is a professor in Ecosystem and Public Health, Faculty of Veterinary Medicine, University of Calgary, but launched her career at the Ontario Veterinary College. Her mixed veterinary practice in Northern Ontario took several turns that resulted in her study of wildlife as sentinels of environmental health or victims of ecosystems in trouble. In the last two decades her major research projects have spanned birds and amphibians on the oil sands of northern Alberta, wolves and bears in coastal British Columbia, storks in southwestern Spain, passerines in the Canary Islands, vultures in South Africa, and most recently the highly endangered greater Northern sage grouse in southeastern Alberta. She has more than 85 peer-reviewed publications. Dr. Smits shares her love of veterinary medicine and ecosystem health research with her graduate and undergraduate students as well as family and friends.

Alastair Summerlee

Alastair James Scott Summerlee, LLD, BSc, BVSc, PhD, MRCVS, President and Vice-Chancellor, University of Guelph (2003-2014). Under his leadership, the University of Guelph was noted for civic engagement and volunteerism. Dr. Summerlee received a 3M teaching fellowship as president. He also continued to teach undergraduate students, supervise graduate students and organize educational student field trips. Dr. Summerlee conducts research in relaxin, cancer biology, iron deficiency anemia in women and HIV/AIDS in aboriginal populations. Dr. Summerlee was chair of the board of the World University Service of Canada for 6 years and became involved in humanitarian issues in Africa, particularly in the refugee camps in Kenya, where he has been working for the last seven years. In the fall of 2012, Dr. Summerlee received the "Award of Highest Honour" from Soka University in Japan, along with the 2012 International Quality of Life Award from Auburn University in the United States.

Sarah Totton

Sarah Totton, DVM, PhD, graduated from the Ontario Veterinary College in 2003 and completed her PhD in Epidemiology (the subject of which is detailed in her story) in 2009. Her short fiction has appeared in magazines and anthologies in North America, the United Kingdom and Australia. In 2007, she was named the Regional Winner (for Canada & the Caribbean) in the Commonwealth Short Story Competition. Her debut short story collection, *Animythical Tales*, featuring animal-themed stories, was published in February 2010 by Fantastic Books (Brooklyn, N.Y.).

David Waltner-Toews

David Waltner-Toews, DVM, PhD, is the author or coauthor of seventeen books of poetry, fiction, recipes and nonfiction. His most recent book is *The Origin of Feces: What Excrement Tells Us About Evolution, Ecology, and a Sustainable Society* (ECW Press, 2013). A University Professor Emeritus at University of

Guelph, he was founding president of Veterinarians without Borders/
Vétérinaires sans Frontières – Canada (www.vwb-vsf.ca) and of the Network
for Ecosystem Sustainability and Health (www.nesh.ca), and a founding
member of Communities of Practice for Ecosystem Approaches to Health in
Canada (www.copeh-canada.org). More information:
www.davidwaltnertoews.com

Scott Weese

Scott Weese, DVM, DVSc, is an associate professor and Canada Research
Chair in the Department of Pathobiology at the Ontario Veterinary College at
the University of Guelph. He has Doctor of Veterinary Medicine and Doctor
of Veterinary Science degrees from the University of Guelph and is an
affiliate faculty member of the Department of Clinical Science at Colorado
State University. Dr. Weese has authored and co-authored an extensive
number of papers.

Acknowledgements

We acknowledge with gratitude the many veterinarians for their contributions to this book. We share their commitment to solving the mysteries of diseases shared between people and animals.

The editorial board of Scott Weese, Jan Sargeant and David Waltner-Toews worked together with Elizabeth Stone and Cate Dewey to select important zoonoses and identify distinguished colleagues, who could tell fascinating stories about the work they do.

We had the assistance of many dedicated people. Skilled project management was done by Tara O'Brien with assistance from Catherine Bianco in the earlier stages, and Joan Alexander in the later stages. Herb Shoveller did initial copy editing and helped some of the authors with their stories. Jane Dawkins brought her creative concepts to the cover design and Char Hoyt created the front cover artwork.

Tony Linka, our illustrator, ably took our words and Samantha Salter's descriptions of life cycles and transformed them into drawings of the sometimes complicated journeys of pests and parasites back and forth between animals, people and the environment.

We are grateful to Karen Richardson for proofreading and her organizing abilities and to Hilde Weisert for book layout and typesetting, and for her constant interest and encouragement.

Index

Ontario Veterinary College

OPUS VETERINUM CIVIBUS

Recent works published by OVC

Live-Love!: Famous Canadians & the Pets they Love

By Nancy Silcox
ISBN: 9780889556201

Nancy Silcox recounts the meaningful memories of notable Canadians and the four-legged friends that have touched their lives. Including: Jann Arden, Robert Bateman, Yannick and Shantelle Bisson, Lynda Boyd , Kurt Browning, The Right Honourable Kim Campbell. Don Cherry, George and Susan Cohon, Mark Cohon, Dr. Stanley Coren , Lynn Crawford , Marc Garneau, Chris and Helene Hadfield, Rick Hansen, Maureen Jennings, Shaun Majumder, Brad Martin and Donna Hayes, Farley and Claire Mowat, David Mirvish, Susan Musgrave, Victoria Nolan, Chantal Petitclerc, Valerie Pringle, Joannie Rochette , Terry Shevchenko, Graeme Smith , Allan Slaight and Emmanuelle Gattuso, Mary Walsh, Brian and Geraldine Williams and Liisa Winkler. Plus a poem by Margaret Atwood.

Animal Companions, Animal Doctors, Animal People

Edited by Hilde Weisert and Elizabeth Arnold Stone
ISBN: 9780889555983

An anthology of poems, stories, essays blending a mix of writers including poets Lorna Crozier, Mark Doty, Patrick Lane, and Molly Peacock with stories from everyday veterinarians and their clients. This book touches on such topics as the bond connecting veterinarians and animals, the never-ending "jobs" of the animals in our lives, and the role of animals in our imagination.

Milestones: 150 Years of the Ontario Veterinary College

by Lisa M. Cox (Author), Peter D. Conlon (Author), Ian K. Barker (Editor)
ISBN: 9780889556010

A brief synopsis of research achievements, teaching advances, people and events throughout the history of the Ontario Veterinary College (OVC). Learn more about the oldest veterinary college in Canada and the United States and how the OVC helped shape the veterinary profession. The book is filled with many photos, some never before published, and an insightful short description on each milestone.

CPSIA information can be obtained
at www.ICGtesting.com
Printed in the USA
LVOW02s1019280416

485288LV00018B/88/P